Expanding Access t Mental Health Counselors

Evaluation of the TRICARE Demonstration

Lisa S. Meredith, Terri Tanielian, Michael D. Greenberg,
Ana Suaréz, Elizabeth Eiseman

Prepared for the Office of the Secretary of Defense
Approved for public release; distribution unlimited

NATIONAL DEFENSE RESEARCH INSTITUTE and
RAND HEALTH

The research described in this report was sponsored by the Office of the Secretary of Defense (OSD). The research was conducted jointly by RAND Health's Center for Military Health Policy Research and the Forces and Resources Policy Center of the RAND National Defense Research Institute, a federally funded research and development center supported by the OSD, the Joint Staff, the unified commands, and the defense agencies under Contract DASW01-01-C-0004.

The RAND Corporation is a nonprofit research organization providing objective analysis and effective solutions that address the challenges facing the public and private sectors around the world. RAND's publications do not necessarily reflect the opinions of its research clients and sponsors.

RAND® is a registered trademark.

Published 2005 by the RAND Corporation
1776 Main Street, P.O. Box 2138, Santa Monica, CA 90407-2138
1200 South Hayes Street, Arlington, VA 22202-5050
201 North Craig Street, Suite 202, Pittsburgh, PA 15213-1516
RAND URL: http://www.rand.org/
To order RAND documents or to obtain additional information, contact
Distribution Services: Telephone: (310) 451-7002;
Fax: (310) 451-6915; Email: order@rand.org

Preface

The military health system serves roughly nine million eligible beneficiaries, including active duty military personnel and their family members, retired military personnel and their family members, and surviving family members of deceased military personnel. Eligible beneficiaries access health care services through the TRICARE program. Mental health care, as well as other forms of health care under TRICARE, is delivered through the direct care system, which consists of military-owned treatment facilities (clinics and hospitals), and the purchased-care system, which consists of coverage for care rendered in the civilian sector. TRICARE provides coverage for most medically necessary mental health care services, including those delivered in inpatient, outpatient, and partial hospitalization settings by qualified providers.

In response to the National Defense Authorization Act (NDAA) for Fiscal Year 2001 (FY01), the Department of Defense implemented a one-year demonstration project designed to expand access to mental health services by easing TRICARE restrictions on services provided by licensed or certified mental health counselors (LMHCs). Currently, LMHCs must meet several eligibility and administrative requirements to serve as authorized TRICARE providers, including documentation of referral and supervision from a physician. Under the demonstration project, LMHCs who met the TRICARE eligibility requirements were allowed to provide services to covered beneficiaries without referral by physicians or adherence to supervisory requirements.

In the NDAA, Congress requested an evaluation of the demonstration's impact on utilization, costs, and patient outcomes. This report describes the evaluation efforts by the RAND Corporation and presents findings based on several sources of data. The report is organized according to specific responses to the evaluation's objectives outlined in the FY01 NDAA and is intended to be included in the sponsor's final report to Congress. The findings may also be of interest to health policymakers in the Department of Defense and mental health policymakers more broadly.

This study was sponsored by the TRICARE Management Activity and was carried out jointly by the Center for Military Health Policy Research, a joint endeavor of RAND Health and the RAND National Defense Institute (NDRI). NDRI is a federally funded research and development center sponsored by the Office of the Secretary of Defense, the Joint Staff, the unified commands, and the defense agencies.

Comments on this report are welcome and may be addressed to the principal investigators, Lisa Meredith (Lisa_Meredith@rand.org; 310-393-0411, ext. 7365) and Terri Tanielian (territ@rand.org; 703-413-1100, ext. 5404). For more information on the Forces and Resources Policy Center, contact the director, Susan Everingham (Susan_Everingham@rand.org; 310-393-0411, ext. 7654). For more information on the RAND

Center for Military Health Policy Research, contact the co-directors, Susan Hosek (sue@rand.org; 310-393-0411, ext. 7255) and Terri Tanielian. They may also be reached by mail at the RAND Corporation, 1776 Main Street, P.O. Box 2138, Santa Monica, CA 90407-2138. More information about RAND is available at www.rand.org.

Contents

Figures

Tables

Summary

The U.S. Congress, in the National Defense Authorization Act (NDAA) for Fiscal Year 2001 (FY01),[1] specified a series of evaluation objectives in requiring a demonstration project designed to expand access to mental health services by easing restrictions on services provided by licensed or certified mental health counselors (LMHCs). The following list provides a preliminary summary of those objectives and RAND's findings on the evaluation.

Legislative-Directed Objectives of This Study and Findings

- **Describe the extent to which expenditures for LMHCs changed as a result of allowing independent practice.** Allowing for increased access to LMHCs had no measurable impact on expenditures for those who received care from LMHCs.
- **Provide data on utilization and reimbursement for non-physician mental health professionals.** Opening up access to LMHCs may have created a small substitution effect—that is, beneficiaries in the demonstration areas were less likely to see other non-physician mental health care providers, such as psychologists, social workers, and psychiatric nurse practitioners. Expenditures for care for those who sought care from non-physician mental health providers significantly increased in both the two demonstration areas and three non-demonstration catchment areas.
- **Provide data on utilization and reimbursement for physicians who make referrals to and supervise LMHCs.** Removing the referral and supervision requirements significantly decreased the likelihood that beneficiaries in the demonstration areas would seek mental health care from a psychiatrist or non-psychiatric physician. There was also a decreased likelihood that beneficiaries in the demonstration areas would receive a psychotropic medication. Expenditures for mental health (MH) care for those who saw physicians increased in both the demonstration and non-demonstration areas, but only the increase for the non-demonstration, non-psychiatric physician group was significant.
- **Describe the administrative costs incurred as a result of documenting referral and supervision.** While difficult to quantify, the demonstration might have resulted in modest cost savings to LMHCs in terms of reduced time and administrative burden, as revealed from our interviews. However, any savings to TRICARE's managed care

---

[1] P.L. 106-398, approved October 30, 2000, 114 Stat. 1654.

support contractors (MCSCs)[2] depended on their baseline enforcement procedures regarding supervision and referral (which were minimal in some cases).

- **Describe the ways in which independent practice authority affects the confidentiality of mental health and substance abuse services for TRICARE beneficiaries.** There was no evidence that independent reimbursement of LMHCs had any impact on patient confidentiality, given that the requirements for supervision and referral do not impact or contradict the standards for confidentiality set forth by the Health Insurance Portability and Accountability Act (HIPAA) of 1996.

- **Describe the effect of changing reimbursement policies on the health and treatment of TRICARE beneficiaries.** Using our survey data, we found no effect on perceived access to mental health services, no effect on self-reported adherence to treatment, and no effect on self-reported mental health status. We found that survey respondents in the demonstration areas reported greater satisfaction with mental health services; however, it is not possible to assess whether the demonstration created the greater satisfaction or if it existed prior to the demonstration.

- **Describe the effect of DoD policies on the willingness of LMHCs to participate as health care providers in TRICARE.** Representatives from the American Counseling Association (ACA) and the American Mental Health Counselors Association (AMHCA) indicated that the practice authority for LMHCs was a disincentive or barrier to LMHCs' participation in the TRICARE network prior to the demonstration. LMHCs in the demonstration and non-demonstration areas said that they view the physician referral and supervision requirement as a potential barrier for patients rather than a source of administrative burden per se. In the demonstration areas, the change in practice authority may have been a motivator for network participation. Enrollment of LMHCs as networked providers increased slightly; however, there were no data to compare this increase with the enrollment of LMHCs in the non-demonstration areas.

- **Identify any policy requests or recommendation regarding LMHCs made by TRICARE plans or managed care organizations.** Based on interviews with representatives from TRICARE MCSCs and TRICARE staff, many MCSCs and TRICARE staff members believe that the adoption of formal standardized training and credentialing requirements could help to facilitate independent practice for LMHCs and could address any concerns about quality of care provided by LMHCs.

Study Background

TRICARE, the program through which beneficiaries of the military health system access health care services, provides coverage for most medically necessary mental health care delivered by qualified providers. The NDAA for FY01 required the Department of Defense (DoD) to conduct a demonstration project involving expanded access under TRICARE to a

---

[2] At the time of this study, TRICARE benefits and coverage policies were implemented through MCSCs (managed care companies under contract with TRICARE to manage and implement TRICARE). They cover 12 geographical regions within the United States.

particular type of mental health service provider—the licensed or certified mental health counselor.

Currently, LMHCs must meet several eligibility and administrative requirements to serve as authorized TRICARE providers. The administrative requirements include documentation of a referral from a physician for each new clinical case and ongoing physician supervision of LMHC services. According to the NDAA, the Secretary of Defense was to conduct a demonstration under which LMHCs who meet eligibility requirements for providers under the TRICARE program may provide services to covered beneficiaries under Title 10 of the U.S. Code without referral by physicians or adherence to existing supervision requirements.

When stipulating the parameters of the demonstration, Congress also required DoD to conduct an evaluation of the demonstration's impact on the utilization, costs, and outcomes of health care services. DoD requested RAND to conduct this evaluation and supply the analyses needed to respond to the evaluation objectives set forth by Congress. This report describes and presents findings from RAND's evaluation.

Under TRICARE, several provider groups are authorized to provide mental health services to beneficiaries, assuming that the individual providers meet eligibility requirements established by TRICARE. The eligible provider groups include physicians, clinical psychologists, clinical social workers, psychiatric nurse specialists, marriage and family therapists, pastoral counselors, and LMHCs. For each provider group, TRICARE stipulates minimum certification or licensure requirements that are relevant to the provider's profession (see *TRICARE Policy Manual 6010.54*, 2002).

As stated above, to be a TRICARE authorized provider, an LMHC must meet several eligibility criteria with respect to training and administrative requirements for his or her practice. The administrative requirements for an LMHC to practice under TRICARE include documentation of a referral from a physician and ongoing supervision of the LMHC's services by a physician. However, services provided by other mental health professionals, including licensed clinical social workers, clinical psychologists, and psychiatric nurse specialists, are currently reimbursed independent of referral or supervision by a physician. TRICARE placed the additional eligibility requirements on LMHCs because of concerns stemming from the lack of nationwide certification standards for this group of mental health professionals.

The professional organizations that represent LMHCs have expressed their concern to the TRICARE Management Activity (TMA; the office within DoD charged with managing TRICARE) and Congress that the eligibility and practice restrictions placed on LMHCs by TRICARE may unduly restrict access to care or may lead potential clients to avoid seeking needed care.

The Demonstration Project

TMA chose to conduct the demonstration project in the Colorado Springs (Ft. Carson and U.S. Air Force Academy) and Omaha (Offutt Air Force Base [AFB]) catchment areas within the TRICARE Central Region. TMA chose these areas because their high volume of LMHCs would ensure ample providers for the demonstration. For purposes of comparison, three non-demonstration catchment areas were chosen: Wright Patterson AFB, Luke AFB, and Ft. Hood. Similar data were collected for beneficiaries in both the demonstration and non-demonstration areas.

Beginning in 2002, Merit-Magellan Behavioral Health, the managed behavioral health care carve-out company for TRIWest Healthcare Alliance,[3] worked collaboratively with TMA to design and implement the demonstration. To advertise the demonstration opportunity, TriWest used a mass mailing to approximately 230 LMHCs who practiced in these areas. LMHCs were informed that by participating in the demonstration, they were eligible to treat TRICARE beneficiaries, over the age of 18 years, without referral or supervision from a physician. To participate, LMHCs were required to sign and return a document titled "Participation Agreement for the TRICARE Expanded Access to Mental Health Counselors Demonstration Project." By signing the participation agreement, counselors agreed to collect a TRICARE Mental Health Counselor Demonstration Project Informed Consent Form (see Appendix A) from each TRICARE patient seen during the demonstration. TRIWest began enrolling LMHCs into the demonstration in late 2002 in preparation for a January 1, 2003, start date. The total number of LMHCs who participated in the demonstration was 123. The relatively low participation rate (55 percent of those who received the mailing) was likely due to the use of only one mass mailing as a means of advertisement.

Evaluation Methods

Our evaluation was guided by a set of general hypotheses based on Avedis Donabedian's model of structure, process, and outcomes of health care (Donabedian, 1980). Accordingly, we expected that the demonstration, which allowed for independent practice by LMHCs, might affect beneficiaries and providers in the following ways: increased access to care delivered by LMHCs (as measured by the percentage of eligible beneficiaries who receive care from LMHCs), higher utilization of mental health services among the eligible beneficiary population in the demonstration areas, decreased total costs of mental health care, and either increased or decreased quality of care.

In the context of this conceptual framework and the evaluation objectives defined by Congress, the purpose of our evaluation analyses was to examine and compare utilization, costs of care, and outcomes for adult beneficiaries receiving mental health services from LMHCs and compare those findings to the findings on beneficiaries seeking services from other mental health providers (including physicians, clinical psychologists, clinical social workers, and others).

To assess the extent to which independent reimbursement of LMHCs affected service utilization, reimbursement costs, and treatment processes, we conducted *secondary analyses of service claims for covered beneficiaries* who received services from mental health providers. These analyses employed a pre-post intervention evaluation methodology that allowed for the identification of any changes over the one-year implementation period among covered beneficiaries in the demonstration catchment areas versus those in the non-demonstration catchment areas.

To assess the impact on treatment and clinical outcomes, we collected and analyzed *primary survey data from a sample of beneficiaries* who received mental health services in the demonstration areas as well as the non-demonstration control areas. These analyses were limited by the requested cross-sectional design; thus, they allow for comparisons between re-

---

[3] TriWest Healthcare Alliance is a management service organization and DoD MCSC. It is one of several private organizations that administer the TRICARE program in various regions of the United States and abroad (see Triwest.com). It is the MCSC responsible for the TRICARE network in the demonstration areas.

spondents in the demonstration areas and respondents in the non-demonstration catchment areas one year post-implementation, but they do not allow for a pre-post evaluation.

We also used *semi-structured qualitative interviewing* techniques to gather relevant information from mental health care providers and MCSCs before and after the implementation of the expanded access demonstration. We used these techniques to determine the administrative costs associated with the documentation of referral and supervision and to assess the impact of independent reimbursement (provided by the demonstration) on a provider's willingness to participate in TRICARE.

We aimed to use both qualitative and quantitative data for this evaluation for several reasons. The type and source of data were typically driven by the nature of the evaluation question and our knowledge of the available and accessible data for responding. We provide additional details on our methodology in Appendix B.

Challenges Associated with the Evaluation

In late 2002, as DoD moved forward with efforts to implement this demonstration and we developed our evaluation strategy, the United States began major deployments in preparation for Operation Iraqi Freedom. At the same time, military personnel were still deployed in Afghanistan for Operation Enduring Freedom. Major combat operations in Iraq began in spring 2003, just as the expanded access demonstration was getting under way. Both the demonstration catchment areas as well as the non-demonstration areas include military installations with deployable forces, both active duty as well as reserve components. While detailed data about the number of personnel deployed from these areas were not available to us, forces were deployed from both the non-demonstration and the demonstration areas during the course of this study.

In an attempt to examine the potential impact of the Iraq war on mental health service needs and utilization, we included items on the beneficiaries' survey that were aimed at eliciting information relevant to this issue. We then aimed to use those data in our multi-variable models to examine differences in self-reported mental health care need, barriers to access, and service utilization between respondents in the demonstration and non-demonstration areas. Because the survey data could not be linked to the administrative claims data, and because there were no comparable administrative data available to us to indicate whether a particular beneficiary had a deployed family member or close friend, we could not examine or control for the impact of the war in the administrative analyses of utilization and costs. Therefore, we offer caution here and again with describing the results that any increases in utilization and costs observed between the pre- and post period in either the demonstration areas or non-demonstration areas could be a consequence of the war in Iraq and not just the demonstration.

Study Results

The Beneficiary Population

Overall, the survey respondent sample was evenly distributed across age groups (14 percent to 23 percent per age group) and was predominantly female (82 percent). Nearly a third had a college education (27 percent), 81 percent were white, and 10 percent were African-American. The majority of the survey respondents were U.S. born (89 percent) and had children (80 percent). Of those with children, 24 percent reported that their children had also received mental health counseling in the past six months. Only 12 percent lived alone, and about half (44.9 percent) were currently working. A fifth of the survey respondents (20 percent) reported that they were not currently working due to health problems. Several demographic differences were noted between the demonstration and non-demonstration re-spondent populations: Respondents in the demonstration areas were younger, more likely to be college educated, less likely to be African-American and more likely to be white, less likely to live alone, and more likely to be currently working compared with beneficiaries in the non-demonstration areas. It should be noted that these differences exist among beneficiaries who use mental health (MH) services as well as those who do not, and likely reflect the dif-ferences associated with these catchment areas. For example, the student population at the U.S. Air Force Academy would likely influence the age distribution in the demonstration areas that includes that catchment area. Several differences were also noted in use of mental health services. Few beneficiaries in the study areas reported awareness of the demonstration.

Beneficiary Outcomes

Little effect of the demonstration was observed on beneficiary outcomes. With two excep-tions, no differences by demonstration area were found in measures of access to mental health services, adherence to treatment, or mental health status: Beneficiaries living in the demonstration areas (regardless of MH provider type) had a 36 percent greater chance of re-porting emotional problems that affected their functioning, but a 32 percent lower likeli-hood of reporting that they had received counseling from a mental health provider in the past six months.

A number of differences between the demonstration and non-demonstration areas were found on Health Plan Employer Data and Information Set (HEDIS) indicators[4] of mental health services. Being in the demonstration areas was associated with greater odds of favorably rating counseling and treatment, a greater chance of reporting an ability to "usually or always" get urgent treatment as soon as needed, greater odds of being able to "usually or always" get an appointment as soon as desired, a greater chance of reporting the ability to get help by telephone, and a lower risk of never having to wait 15 minutes or more to see a clinician.

Other factors associated with access to mental health care include age group, per-ceived barriers to care, a perceived on-the-job stigma from receiving care, and whether a close relative or acquaintance of the beneficiary was deployed to the war in Iraq. Beneficiaries un-der the age of 25 and those who perceived that seeking care would cause them to be stigma-tized at the workplace were less likely to report seeking mental health services. Those who perceived that a stigma from seeking care was a barrier to care were more likely to be taking a prescription medication for a mental health problem. Deployment of a friend or relative was associated with a higher likelihood of receiving counseling from a mental health provider and a lower likelihood of receiving prescription medications for a mental health problem.

Patient confidentiality did not appear to be affected in any way by the demonstra-tion, based on the findings from the beneficiary surveys and provider interviews.

---

[4] This set of indicators is used to rate the quality of services provided by health plans and providers.

Impact on Providers

Interviews with LMHCs were conducted prior to and following the demonstration to assess their attitudes toward the administrative burden of the referral and supervision requirements and their perceptions of the impact of those requirements on beneficiary access to services. Prior to the demonstration, LMHCs tended to regard the referral requirements as a discriminatory policy that reduced access to their services, rather than as a source of administrative burden or increased practice costs. After the demonstration, participating counselors noted that the demonstration had reduced the time needed to obtain referrals. The theme that emerged from the interviews regarding supervision was that baseline supervision practices under TRICARE are highly varied, that some counselors are deeply committed to obtaining supervision regardless of TRICARE's requirements, and that compliance with the supervision requirement was more of a formality than a valuable exercise. Follow-up interviews with providers revealed that removal of the supervision requirement during the demonstration was not perceived as having a major effect on their practice.

Changes in perceptions of professional roles and activities were also assessed. Following the demonstration, LMHCs indicated no demonstration-related changes in their professional roles and activities, apart from reducing the administrative time they spend seeking physician referrals. The primary effect of the demonstration, as perceived by LMHCs, was facilitated access to treatment for TRICARE beneficiaries. The perceptions of other types of MH providers regarding supervision and the scope of LMHC functions were mixed.

Demonstration enrollment records showed that the number of LMHCs who participated in the demonstration increased during the first few months of the demonstration but leveled out during the middle of the demonstration period (likely due to the fact that TMA relied on just the single mailing to advertise the demonstration opportunity). During the demonstration period, the number of LMHCs who enrolled in the TRICARE network increased steadily and modestly in both demonstration areas. Unfortunately, no data were available on the number of enrolled LMHCs in the non-demonstration areas. Thus, it was not possible to assess the role of the demonstration on TRICARE network enrollment.

Impact on TRICARE

The study assessed changes in utilization of mental health services over the demonstration period and endeavored to quantify administrative costs associated with those changes. While controlling for differences in the demonstration and non-demonstration populations, beneficiaries in the demonstration areas were significantly less likely in the post-demonstration period to see a mental health provider other than an LMHC or a psychiatrist, were less likely to see a non-psychiatric physician (such as a primary care physician) for mental health care, and were more likely to have an inpatient psychiatric hospitalization. In addition, we found that those who saw LMHCs in the demonstration areas were significantly less likely to see a psychiatrist or to receive a prescription for a psychotropic medication than those seeing LMHCs in the non-demonstration areas. Based on the administrative nature of the data used to identify these changes, which generally lack clinical information about symptom severity, it was not possible to determine whether the lesser likelihood of seeing a psychiatrist or receiving a psychotropic medication had any clinical significance for this population. That is, it is not possible to determine whether a beneficiary's clinical condition warranted his or her receiving medication and/or psychiatric treatment; however, as a result of the demonstration, there was a lesser likelihood of beneficiaries receiving such treatment.

Changes in patient costs associated with the changes in service utilization were minimal. Attempts to quantify administrative costs associated with referral and supervision and the impact of changes in these policies raised the question of the source of such costs and who, in fact, bears the costs. Costs associated with paperwork would be expected to fall on LMHCs, whereas costs associated with supervision would be expected to fall on the supervising physician; however, neither can be billed to TRICARE. Yet another potential administrative cost associated with supervision and referral is the cost generated from greater demand for and utilization of higher-cost mental health providers, which may result from disincentives to seeking care from LMHCs. To assess the burden of administrative costs to TRICARE, the researchers interviewed representatives from the managed care support contractors (MCSCs) that administered benefits for the demonstration and non-demonstration areas. The consistent theme that emerged from these interviews was that the advantage of the demonstration was not in reducing administrative costs to MCSCs but in increasing access to therapy services for TRICARE beneficiaries. The likelihood that barriers to seeking services from LMHCs would lead beneficiaries to seek care from other, potentially more costly, providers was cited.

Regarding the issue of quality of care, the MCSCs were asked to assess the potential effect on quality of allowing LMHCs greater autonomy. While respondents were divided on whether quality of care might be affected, they agreed that improving credentialing standards for LMHCs, such as through the use of a standardized curriculum, would be a more effective way to promote quality of care and safeguard beneficiaries who seek mental health care.

Conclusions

In summary, our evaluation of the DoD Mental Health Counselor Demonstration for expanded access to mental health counselors under TRICARE found that the demonstration had minimal impact with respect to the variety of outcomes studied here. There were no key effects on expenditures, reimbursement, administrative costs, or patient confidentiality. While we did see increases in utilization and costs for mental health care over the demonstration period, these increases could not be attributed to allowing independent practice authority. Using the administrative data, we found evidence suggesting that the demonstration did affect the type of providers from whom beneficiaries sought mental health care and the likelihood of beneficiaries receiving a psychotropic medication. After controlling for differences in the characteristics of those who see LMHCs, our results revealed a significant decrease in the likelihood of beneficiaries seeing a psychiatrist and a decrease in the likelihood of their receiving a psychotropic drug in the demonstration areas. However, based on administrative data alone, it is not possible to determine whether these changes had a clinically significant impact on beneficiaries.

Where we did observe changes in ratings of satisfaction related to the demonstration, the results were mostly positive. According to self-reported survey data from beneficiaries, those living in the demonstration areas had higher ratings of mental health services.

The effects on administrative costs associated with the requirements for LMHCs were also unclear. From our interviews with LMHCs and other MH providers, it is apparent that supervision and referral were not that onerous to begin with and that any administrative costs associated with the requirements were in fact minimal at the outset. Taken as a whole, our findings suggest that the impact on beneficiaries, providers, and the TRICARE program from expanding access to LMHCs for providers and beneficiaries was minimal.

Interviews with representatives from two of the national counseling associations we surveyed revealed that removal of the referral and supervision requirements for LMHCs remains a top legislative agenda item. Although the ACA and AMHCA have been able to garner the support of some beneficiary advocacy groups, neither the Senate nor House Armed Services Committee staff members whom we interviewed for this study indicated that any other official requests for policy changes had been submitted by beneficiary groups during the most recent session of Congress.

Table S.1 summarizes the key findings and implications for each of the nine legislative objectives for this evaluation that were mandated by Congress. The findings from this demonstration are important in that they show that merely lifting administrative requirements for the provision of mental health care, by itself, is unlikely to result in expanded access and utilization, especially when beneficiaries already have access to other types of mental health providers who do not have the same administrative requirements as the LMHCs but can provide many similar services. Therefore, if the motivation of this demonstration was to reduce the stigma associated with seeking mental health care and to expand access to mental health care services for the military beneficiary population, our findings suggest that efforts in that direction need to go beyond merely lifting the administrative requirements on LMHCs.

Table S.1
Summary of Evaluation Findings and Implications for Each Legislative Objective

Legislation Objective	Key Findings	Implications
1. Describe the extent to which expenditures for LMHCs changed as a result of allowing independent practice	Controlling for beneficiary characteristics, there was no significant change in expenditures for inpatient and outpatient care among the eligible population or among those seeing LMHCs.[a]	Allowing for increased access to LMHCs had no measurable impact on expenditures for mental health services for those who received care from LMHCs.
2. Provide data on utilization and reimbursement for non-physician MH professionals	Among those MH users in the other mental health (OMH) provider group, the mean number of visits increased in both the demonstration and non-demonstration areas.[a] For those in the OMH group, total expenditures for MH care increased in both the demonstration and non-demonstration areas. Comparing the changes pre- and post-demonstration and demonstration versus non-demonstration, we found a decrease in the likelihood of beneficiaries seeing an OMH provider in the demonstration areas.	Opening up access to LMHCs may have created a substitution effect—that is, beneficiaries were less likely to see other non-physician mental health providers, such as psychologists, social workers, and psychiatric nurse practitioners.
3. Provide data on utilization and reimbursement for physicians who make referrals to and supervise LMHCs	Among those MH users in the group of users who saw a psychiatrist, there were no significant changes in the mean number of outpatient MH visits in the demonstration areas or the non-demonstration areas.[a] For those MH users in the non-psychiatrist physician group, there was a statistically significant increase in the mean number of outpatient visits in the non-demonstration areas but not the demonstration areas.[a] Mean expenditures for MH care among MH users in the psychiatrist and other physician groups increased from pre-demonstration to post-demonstration in both the demonstration and non-demonstration areas, but only the increase in the non-psychiatrist "other" physician group in the non-demonstration physician area was statistically significant. Comparing the changes pre- versus post-demonstration and demonstration versus non-demonstration, we found a significant decrease in the likelihood of beneficiaries seeing a physician (psychiatrist or other physician) for MH care in the demonstration areas.	Removing the referral and supervision requirements significantly decreased the likelihood that beneficiaries would get MH care from a physician (psychiatrist or other physician) and, as such, decreased the likelihood that they would also get a psychotropic medication to treat their mental illness.
4. Describe administrative costs incurred as a result of documenting referral and supervision	According to the LMHCs we interviewed, eliminating the physician referral requirement saves time previously spent in telephone consultations to obtain supervision, confirm referrals, and authorize therapy.	The demonstration probably resulted in modest cost savings to LMHCs in terms of time and administrative burden. Any savings to MCSCs depended on their baseline enforcement procedures regarding supervision and referral (which was minimal in some cases).

Table S.1—Continued

Legislation Objective	Key Findings	Implications
5. Compare effect for items outlined in objectives one through four, over one year (pre-post) in the demonstration areas as compared with non-demonstration areas [b]	All findings listed above are based on analyses that compared data gathered from one year prior to the demonstration with data gathered one year following the demonstration in both the demonstration and non-demonstration areas.	Not applicable
6. Describe the ways in which independent practice affects the confidentiality of MH and substance abuse services for TRICARE beneficiaries	There was no evidence that eliminating the referral and supervision requirements would change the standards for confidentiality.	Independent reimbursement of LMHCs would have no impact on confidentiality.
7. Describe the effect of changing reimbursement policies on the health and treatment of TRICARE beneficiaries	There was no effect on perceived access to MH services. There was no effect on self-reported adherence to MH treatment. There was no effect on self-reported MH status. There was a potential positive effect on HEDIS ratings of mental health services; however, positive ratings may have also been evident prior to the demonstration.	Increased access to LMHCs had no adverse effect on TRICARE beneficiaries and may be associated with greater satisfaction with MH services.
8. Describe the effect of DoD policies on the willingness of LMHCs to participate as health care providers in TRICARE	Lack of independent practice authority for LMHCs was viewed as a disincentive or barrier to participation prior to the demonstration. Demonstration participation increased initially and leveled off around the middle of the demonstration period. Enrollment of LMHCs as TRICARE network providers increased during the demonstration period, but this is not likely the result of the changing practice authority because this was a temporary demonstration.	The findings suggest that the demonstration may have been a motivator to network participation (although we have no data on network enrollment for the non-demonstration catchment areas during the same time period to use for comparison).
9. Identify any policy requests or recommendations regarding LMHCs made by TRICARE plans or managed care organizations	Removal of the referral and supervision requirements for LMHCs remains a top legislative priority for the ACA and AMHCA. According to MCSC representatives, quality concerns could be addressed by development of appropriate and standardized credentialing mechanisms.	Adoption of formal credentialing standards could help to facilitate independent practice for counselors in states with rigorous licensing, while helping to promote the implementation of similar standards elsewhere.

[a] We created hierarchical groups of users by provider type to compare differences in the changes in users' utilization patterns (see Chapter Five).

[b] Item 5 was included in the legislation as a means of describing the methods to be used for responding to objectives 1 through 4. Although it is not included as an objective in the bulleted list at the top of this summary, we include it here for consistency with the legislation.

Acknowledgments

The authors wish to thank several individuals for their guidance and support in carrying out this work. We are especially grateful to Capt. Mark Paris, Deputy Director of the Clinical Quality Programs Division for the Chief Medical Officer for TMA, who provided valuable feedback on the history of practice authority for mental health providers within DoD, feedback on survey instruments, and insight into the results. We acknowledge the support of the project officer, Col. Merrily McGowan, and Patricia Golson and the staff within TMA's Health Program Analysis and Evaluation office for facilitating this study. We thank the TMA Privacy Office for its help in securing access to the required data for this study. We acknowledge the diligence and support of our project team, Renee Labor, Lisa Jaycox, Greg Ridgeway, Michael Schoenbaum, Barbara Wynn, and Harold Alan Pincus. We would also like to thank Naakesh A. Dewan, President, Center for Mental Healthcare Improvement, Clearwater, Florida, Bradley D. Stein, and Kanika Kapur for their reviews of the draft of this report and their valuable comments. We also thank Sydne Newberry for her help in crafting a previous version of this report, and we are indebted to Nancy DelFavero for her careful edit and help in improving the final report.

Finally, we also thank the military health system beneficiaries who took the time to complete the survey instrument. Without their responses, this report would not have been possible.

Acronyms

ACA	American Counselors Association
AD	active duty
ADD	dependents of active duty (personnel)
AFB	Air Force Base
AMHCA	American Mental Health Counselors Association
CFR	Code of Federal Regulations
CHAMPUS	Civilian Health and Medical Program of the Uniformed Services
CI	confidence interval
CM	Clinical Modification (codes)
CPT	Current Procedural Terminology
CY	calendar year
DEERS	Defense Enrollment Eligibility Reporting System
DoD	Department of Defense
DSM-IV	*Diagnostic and Statistical Manual,* Fourth Edition
ECHO	Experiences of Care and Health Outcomes
F.R.	Federal Register
FY	fiscal year
GAO	Government Accounting Office
HCSR	Health Care Service Record
HEDIS	Health and Employer Data and Information Set
HIPAA	Health Insurance Portability and Accountability Act
ICD	International Classification of Diseases
LMHC	licensed and/or certified mental health counselor
MBC	Merit Behavioral Care
MCSC	managed care support contractor
MH	mental health
MHS	military health system
MOS	Medical Outcomes Study

MTF	military treatment facility
NCQA	National Commission on Quality Assurance
NDAA	National Defense Authorization Act
NDC	National Drug Code
NP	not presented
ns	not (statistically) significant
OMH	other mental health (provider)
OR	odds ratio
PCP	primary care physician
PDTS	Pharmacy Data Transaction Service
PHQ	Patient Health Questionnaire
PIC	Partners In Care
PITE	point-in-time extract
P.L.	Public Law
RDD	dependents of retirees
Rx	prescription
SAS	Statistical Analysis Software
SE	standard error
SRG	(RAND) Survey Research Group
TMA	TRICARE Management Activity
TMOP	TRICARE Mail Order Pharmacy
TPO	treatment, payment, and operations
USAF	U.S. Air Force

CHAPTER ONE

Introduction

TRICARE, the program through which beneficiaries of the military health system access health care services, provides coverage for most medically necessary mental health care delivered by qualified providers. The National Defense Authorization Act (NDAA) for Fiscal Year 2001 (FY01)[1] required the Department of Defense (DoD) to conduct a demonstration project involving expanded access under TRICARE to a particular type of mental health service provider—the licensed or certified mental health counselor (LMHC).

Currently, LMHCs must meet several eligibility and administrative requirements to serve as authorized TRICARE providers. The administrative requirements include documentation of a referral from a physician for each new clinical case and ongoing physician supervision of LMHC services. According to the NDAA, under the demonstration, LMHCs who met eligibility requirements for providers under the TRICARE program could provide services to covered beneficiaries without referral by physicians or adherence to supervision requirements.

When stipulating the parameters for the demonstration, Congress also required DoD to conduct an evaluation of the demonstration's impact on the utilization, costs, and outcomes of health care services. DoD asked RAND to conduct this evaluation and the analyses required to respond to the evaluation objectives set forth by Congress. (These objectives are outlined in greater detail in Chapter Two.) This report describes and presents findings from RAND's evaluation.

In this introductory chapter, we provide a brief overview of the TRICARE program, describe TRICARE's coverage for mental health services and policies regarding providers, and discuss the motivation for the demonstration.

Background on TRICARE

The TRICARE program was established in 1992 to reorganize the Civilian Health and Medical Program of the Uniformed Services (CHAMPUS). TRICARE created a comprehensive managed health care program for the delivery and financing of health care services in the military health system (MHS). Entitlement to TRICARE benefits is set forth and defined in Title 10 of the U. S. Code and generally includes all active duty personnel and military retirees and their eligible dependents. With a few exceptions, identified in Title 10,

---

[1] P.L. 106-398, approved October 30, 2000, 114 Stat. 1654.

those eligible for TRICARE must be listed in the Defense Enrollment Eligibility Reporting System (DEERS) in order to receive care.[2]

In fiscal year 2003 (FY03), it was estimated that approximately 9.1 million individuals were eligible for benefits within the military health system, including approximately 1.87 million active duty personnel, 2.45 million family members of active duty personnel, and 4.76 million retirees and their family members. This estimate represents an increase from prior fiscal years (8.4 million in FY01 and 8.7 million in FY02), largely due to the mobilization of large numbers of National Guard and Reserve members and the extension of health benefits to their family members (Institute for Defense Analyses et al., 2004).

For military beneficiaries under age 65, TRICARE offers several options for care: TRICARE Prime, Standard, and Extra. TRICARE Prime is essentially a health maintenance organization; the provider network consists primarily of military treatment facilities (MTFs) (the "direct care" system), supplemented by care from designated civilian providers as authorized (the "purchased care" system). Active duty personnel are automatically enrolled in TRICARE Prime. Non–active duty beneficiaries (e.g., family members) who enroll in TRICARE Prime receive priority access to care at MTFs and are required to follow the referral and utilization management guidance of a primary care manager.

In FY03, roughly 67 percent of all eligible beneficiaries were enrolled in TRICARE Prime (Institute for Defense Analyses et al., 2004). Beneficiaries who do not enroll in TRICARE Prime are automatically eligible for TRICARE Standard or Extra; these beneficiaries remain eligible for MTF care on a space-available basis, with low priority. TRICARE Standard and Extra function essentially as a preferred provider organization. ("TRICARE Extra" refers to the covered use of in-network providers; "TRICARE Standard" refers to the covered use of out-of-network providers.) During FY03, nearly 75 percent of all eligible beneficiaries under the age of 65 used at least one MHS service from either a direct or purchased source of care. So, while there are close to nine million eligible MHS beneficiaries, approximately 6.75 million use the MHS (Institute for Defense Analyses et al., 2004).

TRICARE Coverage Policies

TRICARE coverage policies are set forth in 32 Code of Federal Regulations (CFR) Part 199. The TRICARE Management Activity (TMA) (as delegated by the Assistant Secretary of Defense for Health Affairs) has authority for developing policies and regulations required to administer and manage the TRICARE program effectively. Basic coverage in TRICARE's programs includes most medically necessary care rendered to beneficiaries by authorized providers. Benefits include specified medical services and supplies from authorized civilian sources such as hospitals, other authorized institutional providers (e.g., residential treatment centers), physicians, other authorized individual professional providers (nurse practitioners, physician assistants, clinical social workers), and professional ambulance services, prescription drugs, authorized medical supplies, and rental or purchase of durable medical equipment.

---

[2] The exceptions include Medal of Honor recipients and eligible dependents, NATO dependents in the United States on a peacekeeping mission, abused dependents of discharged active duty personnel, and newborns entered into the DEERS system within the year of their birth.

Detailed definitions, inclusions and exclusions, and requirements for coverage are outlined in 32 CFR Part 199.4

At the time this research was conducted, TRICARE benefits and coverage policies were implemented through TRICARE Managed Care Support Contractors (MCSCs) covering 12 geographic health care regions within the United States.[3] TRICARE's Quality and Utilization Review Peer Review Organization Program assists in monitoring utilization, reviewing claims, and considering appeals for coverage. At present, TRICARE claims are processed by private claims-processing contractors.

Currently, TRICARE covers most treatments for most conditions; however, the statute governing TRICARE prohibits treatment for smoking cessation and weight management and restricts inpatient psychiatric care to 30 days per fiscal year for adults. TRICARE covers 80 percent of most outpatient mental health services (including psychotherapy) provided by qualified providers but imposes some restrictions on the frequency and length of visits to be covered (e.g., preauthorization is required for more than eight psychotherapy visits, and coverage is limited to 60 visits for substance abuse treatment in a benefit period).[4] Up to eight additional psychotherapy visits can be preauthorized per request if deemed necessary by the contractor. However, some variation exists among MCSCs in how these visits are preauthorized.

TRICARE also provides beneficiaries with pharmacy benefits: TRICARE beneficiaries incur nominal copayments for medications depending on the type of drug (generic versus brand-name) and the mode of prescription fulfillment (MTF, the TRICARE Mail Order Pharmacy [TMOP] program, network retail pharmacy, or non-network retail pharmacy).

Practice Authority for Mental Health Care Providers

Under TRICARE, several provider groups are authorized to provide mental health services to beneficiaries, assuming the individual providers meet eligibility requirements established by TRICARE. The eligible provider groups include psychiatrists as well as non-psychiatric physicians, clinical psychologists, clinical social workers, psychiatric nurse specialists, marriage and family therapists, pastoral counselors, and mental health counselors. For each provider group, TRICARE stipulates minimum certification or licensure requirements as relevant to the profession (see *TRICARE Policy Manual 6010.54, 2002*).

As stated above, LMHCs must meet several eligibility and administrative requirements to be an authorized TRICARE provider. The eligibility requirements for LMHCs are similar to those stipulated for clinical social workers. They include the following:

- a master's degree in mental health counseling or an allied mental health field from a regionally accredited institution
- two years of post-master's experience to include 3,000 hours of clinical work and 100 hours of face-to-face supervision

---

[3] As of November 1, 2004, the 12 geographic regions had been condensed into four regions.

[4] A benefit period is defined as 12 months, or one year.

- licensure or certification as a mental health counselor; if a jurisdiction does not offer licensure/certification, the counselor must be (or meet all requirements to become) a Certified Clinical Mental Health Counselor as determined by the National Board of Certified Counselors.

The administrative requirements for LMHCs to practice under TRICARE include documentation of a referral from a physician and ongoing supervision of LMHCs' services by a physician. However, services provided by other mental health professionals, including licensed clinical social workers, clinical psychologists, and psychiatric nurse specialists, are currently reimbursed independent of referral or supervision by a physician.

Motivation and Impetus for the Demonstration

Ensuring TRICARE beneficiaries' access to quality mental health care is critically important. Beneficiaries are typically family members of active duty military members or retired service personnel who depend on the TRICARE health plan for all or nearly all of their health care. Military families, as compared with most civilian families, are subject to unique forms of stress, such as deployments of the service member (often to sites of extreme danger), deployment or security alerts, and frequent relocation (Orasanu and Backer, 1996), all of which can be disruptive and may, in some cases, precipitate new mental health problems or exacerbate existing ones. For example, from recent DoD Surveys of Health Related Behaviors Among Military Personnel, we know that the majority of active duty personnel report "some" to "a lot" of stress associated with their work, with deployment and separation from their family being listed as the most frequently indicated stressors (Bray et al., 2003). Bray et al. also reported a substantial prevalence of symptoms of anxiety and depression (16.6 percent and 18.8 percent, respectively) among active duty personnel. Yet, while 19 percent of the personnel responding to Bray et al.'s survey reported a need for mental health care, only about two-thirds of them reported receiving this care.

The risk of mental health problems and the need for mental health services are greater during wars and conflicts. The results of a recent study published in the *New England Journal of Medicine* (Hoge et al., 2004) showed that overseas deployment increased the rate of mental disorders among service personnel. The Hoge et al. study of Army and Marine Corps personnel serving in Iraq and Afghanistan found that a higher percentage of military members were at risk for mental illness after deployment than prior to deployment. Only a small proportion of those experiencing symptoms sought mental health care, and other respondents cited the perceived stigma of seeking mental health care as a key barrier. Army troops in Iraq have also been reported to have a significantly higher rate of suicide than the general population (Dunnigan, 2004). Because a large proportion of these military personnel are married and many have children, the potential consequences for spouses and children must be considered. All of these factors (i.e., stress associated with work, separation from family, and anxiety and depressive symptoms among military members) can have indirect consequences on the mental health of family members of active duty and former military members.

So while military health system beneficiaries may have a great need for mental health services, studies have indicated that their own concerns about being stigmatized may be a major barrier to their ability to access and receive care. So, why implement this particular demonstration?

During our interviews with congressional staff, representatives from military beneficiary groups, and national professional counseling associations, including the American Counselors Association (ACA) and the American Mental Health Counselors Association (AMHCA), we learned that the legislation mandating the demonstration was developed following requests initiated in 1999 from the professional associations to Congress for a change in practice authority for LMHCs under TRICARE (further discussed in the next section). The associations articulated concerns among their constituents about the referral process creating a barrier to beneficiaries seeking care. They also expressed concerns that the supervision requirement posed an additional administrative cost to the program and created potential problems in regard to professional autonomy and patient confidentiality. They based these concerns on phone calls and other anecdotal reports from their membership. To the best of our knowledge and based on our research, while beneficiaries have expressed concerns about access to TRICARE-eligible providers in general (particularly in rural or remote areas) and to mental health services in particular (Schone, Huskamp, and Williams, 2003), there were no available data indicating specific concerns from beneficiaries about accessing LMHCs.

According to those we interviewed with the ACA and AMHCA, independent practice authority under TRICARE had been granted to clinical social workers, psychiatric nurse practitioners, and marriage and family therapists in the 1980s, and the associations at the time said that their members wanted the same opportunities. Representatives from the TMA whom we interviewed, however, indicated that the administrative referral and supervision requirements in place for LMHCs are based on concerns about quality. In an information paper provided to Congress at the time of the implementation of the NDAA legislation and to the study team during our evaluation of the demonstration, TMA noted the lack of a uniform standard curriculum used nationwide to guide the training of LMHCs ("Information Paper: On Independent Practice . . . ," 2000). Further, the paper explained that the purpose of the physician supervision requirement is to ensure that the quality of care provided to TRICARE beneficiaries is not compromised by a provider's scope of training and experience being different from that of other currently authorized groups of providers.

In responding to requests from the professional counseling associations, Congressman Walter Jones (North Carolina) introduced language into the NDAA for FY01 to change the practice authority for LMHCs under TRICARE. Due to concerns about the impact this change might have on health care utilization and costs of mental health care, House Armed Service Committee staff suggested a compromise position and revised the language to include a demonstration and subsequent evaluation of the impact of changing practice authority. The NDAA for FY01, including the required demonstration and evaluation, then became law.

Stakeholder Requests for Changes to TRICARE Policies

During our interviews with stakeholders—including representatives from the ACA and AMHCA professional counseling associations, TRICARE MCSCs, and staff members from the Senate and House Armed Services Committee—we inquired about requests for policy changes with respect to the practice authority of LMHCs.

Removal of the referral and supervision requirements for LMHCs remains a top legislative agenda item for both the AMHCA and ACA. This issue has continuously been among the priority items listed on both associations' Web sites (http://www.amhca.org/policy/ and http://www.counseling.org/AM/Template.cfm?Section=PUBLIC_POLICY) and was repeatedly mentioned during our interviews and subsequent inquiries from the associations with respect to the status of our study. It should be noted that the AMHCA and ACA were able to garner the support of some beneficiary advocacy groups, such as the National Military Family Association, in their original request to seek legislative change in practice authority under TRICARE. However, when we spoke with staff members of the Armed Services Committees in the House and Senate in 2003, they indicated that no other official requests for policy changes to implement independent practice authority for LMHCs, or to expand access to mental health care services within TRICARE more generally, had been submitted by stakeholders.

We should also note that several of the MCSC officials with whom we spoke acknowledged the potential unfairness of current referral and supervision requirements for LMHCs and the perception that these requirements may tend to drive beneficiaries toward other types of providers for their mental health care. The consensus view among the MCSC representatives was that these requirements are not a particularly effective way to promote the quality of care. Instead, the MCSC representatives suggested that concerns regarding quality of care might be addressed more readily through appropriate credentialing mechanisms for counselors, perhaps as national standards for licensure that the TMA could endorse. Adoption by the TMA of formal credentialing standards could facilitate independent practice for counselors in states with rigorous licensing standards, while helping to promote the implementation of similar licensing standards in other parts of the country.

Organization of This Report

In Chapter Two, we provide a description of the demonstration itself, including details on how and where the Department of Defense implemented the program; outline the evaluation objectives; present our conceptual framework for approaching the study; and discuss the research methods we employed. In Chapters Three and Four, we present findings, based on our survey of TRICARE beneficiaries and interviews with other stakeholders in the system, on the demonstration's impact on health care utilization, cost, and outcomes, from the perspective of the beneficiary and the provider, respectively. In Chapter Five, we present our findings on the demonstration's effect on the TRICARE system, with respect to utilization and costs based on administrative claims data. (We recognize that beneficiaries and providers are part of the overall TRICARE system; however, organizing the results in this fashion allowed us to use a conceptual framework, described in Chapter Two, to categorize the objectives of the evaluation and our data sources.) Finally, in Chapter Six, we present our conclusions and discuss the implications and the limitations of our findings. The appendices of this report provide technical documentation of our work.

Evaluating the Impact of the Demonstration: Implementation, Objectives, Framework, and Methods

Implementation of the Demonstration

The TMA chose to conduct the demonstration project in the Colorado Springs (Ft. Carson and U.S. Air Force [USAF] Academy) and Omaha (Offutt Air Force Base [AFB]) catchment areas within the TRICARE Central Region (which consists of New Mexico; Nevada; Arizona, except for the Yuma area; the southwestern corner of Texas, including El Paso; Colorado; Utah; Wyoming; most of Idaho, Montana; North Dakota; South Dakota; Nebraska; Kansas; Minnesota; Iowa; and Missouri, except for the St. Louis area).[1] At the time of the demonstration, the Managed Care Support Contractor in this region was TriWest.[2] Beginning in 2002, Merit-Magellan Behavioral Health, the managed behavioral health care carveout company for TriWest, worked collaboratively with TMA to design and implement the demonstration. Implementation plans called for a mass mailing to approximately 230 LMHCs practicing in these areas to advertise the opportunity to be part of the demonstration. Both LMHCs enrolled in the TRICARE network and those not enrolled (but eligible for enrollment) were eligible for participation. Thus, the mailing was targeted to both enrollees and non-enrollees (additional information used to construct the mailing lists was supplied by the American Counselors Association). LMHCs were informed that by participating in the demonstration, they were eligible to treat TRICARE beneficiaries, over the age of 18 years, without referral or supervision from a physician.

To participate, LMHCs were required to sign and return a document titled "Participation Agreement for the TRICARE Expanded Access to Mental Health Counselors Demonstration Project." By signing this document, LMHCs agreed to collect a TRICARE Mental Health Counselor Demonstration Project Informed Consent Form (see Appendix A) from each TRICARE patient seen during the demonstration. If counselors did not return the agreement to TriWest, they were excluded from the demonstration and were required to comply with the TRICARE physician referral and supervision requirements.

Plans for the demonstration were published in the *Federal Register* (67 FR 57581). TriWest began enrolling LMHCs into the demonstration in late 2002 in preparation for a January 1, 2003, start date.

---

[1] The TRICARE West region includes Alaska, Arizona, California, Colorado, Hawaii, Idaho, Iowa (except 82 Iowa zip codes that are in the Rock Island, Illinois, area), Kansas, Minnesota, Missouri (except the St. Louis area), Montana, Nebraska, Nevada, New Mexico, North Dakota, Oregon, South Dakota, Texas (the southwest corner only, including El Paso) Utah, Washington, and Wyoming.

[2] TriWest Healthcare Alliance is a management service organization and DoD MCSC. It is one of several private organizations that administer the TRICARE program in various regions of the United States and abroad (see Triwest.com). It was the MCSC responsible for the TRICARE network in the demonstration areas.

Beginning in December 2002 and in each subsequent month through December 2003, TriWest submitted a detailed report to TMA on enrollment of LMHC participants. Table 2.1 summarizes the number of participating counselors for each month of the demonstration, as reported by TriWest. The total number of LMHCs who participated in the demonstration was 123.

The demonstration ended on December 31, 2003, at which time LMHC participation was terminated. At the same time, referral and supervision requirements for new patients and for episodes of care were reinstated.[3]

Selection of Non-Demonstration Comparison Sites

As stipulated by legislation (P.L. 106-398), the evaluation of the demonstration's impact would require comparison of utilization, costs, and outcomes of care provided by LMHCs under the demonstration with comparable data for similar areas in which the demonstration was not being implemented. In late 2002, TMA project officers selected three catchment areas to serve as non-demonstration comparison sites for data collection and analyses—Wright-Patterson AFB near Dayton, Ohio; Luke AFB near Phoenix, Arizona; and Ft. Hood near Killeen, Texas. The rationale and criteria used to select these sites are detailed in Appendix B.

Evaluation Objectives

In specifying the objectives of the required evaluation, Congress requested analyses to determine the extent of the demonstration's impact on the utilization, costs, and outcomes of care

Table 2.1
Demonstration Participation by Catchment Area and Month

Month	Colorado Springs	Omaha	Total
January 2003	41	41	82
February 2003	57	53	110
March 2003	62	55	117
April 2003	64	55	119
May 2003	67	55	122
June 2003	68	55	123
July 2003	68	55	123
August 2003	68	55	123
September 2003	67	55	122
October 2003	66	55	121
November 2003	66	55	121
December 2003	66	55	121

---

[3] Participating LMHCs were allowed to continue independent treatment of patients who began treatment before December 31, 2003, and were still within a current episode of authorized care (e.g., first eight therapy visits) without referral or supervision.

provided by LMHCs and other mental health care providers. Congress requested that the final evaluation report include the following:

1. A description of the extent to which expenditures for reimbursement of licensed or certified professional mental health counselors changed as a result of allowing the independent practice of licensed and/or certified mental health counselors
2. Data on utilization and reimbursement regarding non-physician mental health professionals, other than licensed or certified professional mental health counselors, under CHAMPUS and the TRICARE program
3. Data on utilization and reimbursement regarding physicians who make referrals to, and supervise, mental health counselors
4. A description of the administrative costs incurred as a result of the requirement for documentation of referrals to mental health counselors and supervision activities for LMHCs
5. For each of the categories described in paragraphs 1 through 4, a comparison of data for a one-year period for the areas in which the demonstration project is being implemented with corresponding data for a similar areas in which the demonstration project is not being implemented
6. A description of the ways in which allowing for independent reimbursement of licensed or certified professional mental health counselors affects the confidentiality of mental health and substance abuse services for covered beneficiaries under CHAMPUS and the TRICARE program
7. A description of the effect, if any, of changing reimbursement policies on the health and treatment of covered beneficiaries under CHAMPUS and the TRICARE program, including a comparison of the treatment outcomes of covered beneficiaries who receive mental health services from licensed or certified professional mental health counselors acting under physician referral and supervision, other non-physician mental health providers recognized under CHAMPUS and the TRICARE program, and physicians, with treatment outcomes under the demonstration project allowing independent practice of professional counselors on the same basis as other non-physician mental health providers
8. The effect of policies of the Department of Defense on the willingness of licensed or certified professional mental health counselors to participate as health care providers in CHAMPUS and the TRICARE program
9. Any policy requests or recommendations regarding mental health counselors made by health care plans and managed care organizations participating in CHAMPUS or the TRICARE program.

Conceptual Model

Our evaluation was guided by a set of general hypotheses based on Avedis Donabedian's model of structure, process, and outcomes of health care (Donabedian, 1980). Accordingly, we expected that the demonstration, which allowed for independent practice by LMHCs, might have the following effects on beneficiaries and providers:

- Increased *access* to care delivered by mental health counselors resulting from fewer procedural barriers and less of a stigma from seeking counseling services, in contrast with no increased access to psychotropic medication care due to getting medicines solely from a doctor.
- Higher *utilization* of mental health services (especially counseling) as a function of direct access to LMHCs. There may be an increase in beneficiaries receiving both medication and counseling.
- Decreased total *cost* of care, again due to more use of mental health counselors (as a lower-cost alternative to other mental health specialists) and elimination of supervision costs.
- Increased or decreased *quality of care* among those seeing mental health counselors. Increased quality of care could be due to changes in professional roles, including greater autonomy and responsibility, earlier access to care, and earlier interventions. However, the demonstration could decrease quality of care through lower rates of collaboration with other professionals, especially for psychotropic medication treatment in collaboration with physicians, or through inappropriate visits, or based on some characteristics potentially associated with counselors (such as lower use of evidence-based therapy, lack of clinical skill to detect problems).

In Figure 2.1, we illustrate how various mechanisms of change may operate to affect the outcomes listed above. The framework incorporates the interrelated perspectives of two types of stakeholders—beneficiary and provider.

One way to assess the effects of the demonstration is in terms of beneficiaries' access to mental health care and the utilization, cost, and quality of mental health care. The demonstration added several pathways to care by increasing the independence of LMHCs. In particular, under the demonstration, beneficiaries could self-refer directly to an LMHC. By contrast, non-demonstration LMHCs may see only patients who are referred to them by other providers. Self-referral may lead to greater availability of counseling services but would not be expected to change the availability of psychotropic medications. On the other hand, self-referral to LMHCs might change the demand for medications, because the demonstration could result in more people receiving both medication and counseling. From the beneficiary's point of view, seeking care directly from a mental health counselor may carry less of a stigma, because it is no longer necessary to obtain approval for a referral from a physician. This may be particularly true for those individuals not willing to discuss their mental health concerns with primary care providers they see on base.

Another way to examine the effect of the demonstration is to assess its systemic effects on providers (mental health counselors, psychiatrists, other mental health specialists, and primary care physicians) including their perceptions of professional autonomy and role changes. We expect variation in the impact to different provider groups; for example, we anticipate a potential increase in utilization of mental health counselors, but a potential decrease or no change in utilization of other mental health specialists and physicians (including psychiatrists). With the demonstration, we might expect a lesser administrative burden because documentation for referrals and supervision is no longer required. We also expect more

Figure 2.1
Conceptual Framework for Evaluating the Effects of the DoD Mental Health Counselor Demonstration

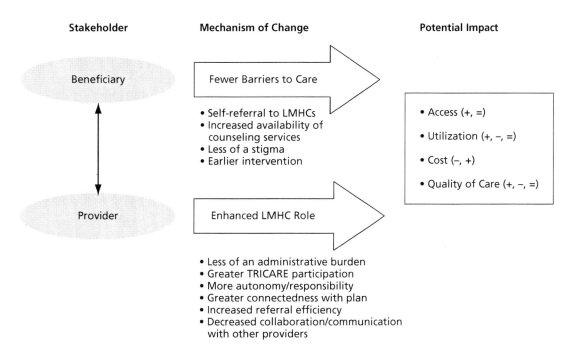

RAND *MG330-2.1*

NOTE: + denotes increase; – denotes decrease; = denotes no change/same.

participation in the TRICARE network by LMHCs. The ability for mental health counselors to practice independently will allow for more professional autonomy and greater responsibility for beneficiaries, and this could potentially lead to better care. In fact, LMHCs may pay greater attention to TRICARE policies and could become more involved in quality improvement activities. The referral process would be more efficient for many of the same reasons noted above. However, a potential negative impact on outcomes could result from less interaction between providers, which might lead to uncoordinated, duplicative, and unsupervised care. The remainder of this report is organized around this framework. We first present results from the perspective of TRICARE beneficiaries followed by those from the providers' perspective. Finally, we present data obtained from administrative records to represent systemic effects.

Evaluation Methodology

In the context of this conceptual framework and the Congressional objectives, the purpose of our evaluation analyses was to examine and compare utilization, costs of care, and outcomes for beneficiaries receiving mental health services from LMHCs and compare such outcomes to beneficiaries seeking services from other mental health providers (including physicians, clinical psychologists, clinical social workers, and psychiatric nurse practitioners).

To assess the extent to which independent reimbursement of LMHCs impacts service utilization, reimbursement costs, and treatment process outcomes, we *conducted secondary analyses of service claims* for covered beneficiaries receiving services from mental health providers. These analyses employed a pre-post intervention evaluation methodology that allows for the identification of any changes over the one-year implementation period among covered beneficiaries in the demonstration areas versus those in the non-demonstration catchment areas.

To assess the impact on treatment and clinical outcomes, we *collected and analyzed primary survey data* from beneficiaries in the demonstration region as well as the non-demonstration control region. These analyses were limited by the requested cross-sectional design; thus, they allow for comparisons between respondents in the demonstration and non-demonstration catchment areas one year post-implementation, but they do not allow for a pre-post evaluation.

We also *used semi-structured qualitative interviewing techniques* to gather relevant information from mental health care providers and managed care organizations (before and after the implementation of the expanded access demonstration) to determine the administrative costs associated with the documentation of referral and supervision and to assess the impact of independent reimbursement on provider willingness to participate in TRICARE.

We aimed to use both qualitative and quantitative data for this evaluation for several reasons. The type and source of data were typically driven by the nature of the evaluation question and our knowledge of the available and accessible data for responding. For example, claims data are best suited for examining utilization and costs, but do not contain any information about satisfaction with or outcomes of care. We believe that combining qualitative and quantitative data and the multiple data sources adds to the breadth of the perspectives we were able to capture for the evaluation. We provide additional details on each of these methodologies in the following subsections and in Appendix B.

Secondary Analysis of Claims Data

To assess the extent to which independent reimbursement of LMHCs impacts service utilization and expenditures, we conducted analyses of service claims for covered beneficiaries receiving services from mental health providers. We compared data on claims for care provided within the demonstration areas to data from pre-selected non-demonstration areas (the comparison areas) using both one year of pre-demonstration data and one year of post-implementation data.

Data Sources. To conduct these analyses, our study relied upon several DoD health care data sets. We requested calendar year (CY) 2002 and 2003 Health Care Service Records and pharmacy records from the Pharmacy Data Transaction Service for TRICARE beneficiaries who received mental health services (broadly defined; see the next paragraph) in the specified catchment areas (demonstration and comparison). We also requested data from DEERS (e.g., the most recent available point-in-time extract [PITE]) so that we could estimate mental health service utilization rates among eligible beneficiaries for each catchment area of interest.

Definition of Mental Health Service User. To ensure comprehensiveness in our sample, we employed a broad definition of mental health service users to include beneficiaries who received TRICARE-covered care, during the one-year period before the implementation of the demonstration or during the one-year period following the implementation of the demonstration and who met one or more of the following criteria:

- Visit to a mental health specialty provider (defined by the provider codes for LMHC, clinical social worker; psychologist, family/marital therapist, or psychiatrist)
- Visit for a mental health service (defined by the physician's Current Procedural Terminology [CPT] code or International Classification of Diseases [ICD] procedural codes for psychotherapy, psychoanalysis, psychiatric management, counseling, or group/family therapy, and other such care)
- Claim for a psychotropic medication prescription (defined by National Drug Codes [NDCs] for psychotropic medication, e.g., antidepressants, stimulants, antipsychotics, anxiolytics, and other such medications)
- A mental health diagnosis (with ICD 9-CM codes 292–312, 314) appearing in one of the diagnosis fields. Beneficiaries with a secondary or tertiary mental health diagnosis were considered mental health service users if only one of the other criteria were met.

Analytic Design. For the majority of these analyses, we employed a pre-post intervention evaluation methodology. Once the data were formatted and prepared for analyses, using the pre-post intervention design, we examined utilization patterns and reimbursement data for a one-year period prior to the demonstration (i.e., the baseline) and a one-year period of data following full implementation of the demonstration. The main evaluation analyses measured changes pre- and post-demonstration in the amount, type, and cost of mental health services provided to TRICARE beneficiaries. All analyses examined group differences between beneficiaries in the demonstration areas and those receiving care in the non-demonstration (comparison) areas and as differences by type of provider (LMHCs, other mental health [OMH] providers, and physicians, which we broke out further into psychiatrists and other non-psychiatrist physicians). Using a hierarchical approach, we grouped providers first by LMHCs, followed by psychiatrists, non-physician OMH providers, then by other physicians (e.g., primary care, internal medicine). We used this hierarchical approach to isolate those beneficiaries who received care from LMHCs as the primary group of interest and then to eliminate overlap among the groups. We do not intend for these hierarchical groups to be directly comparable to one another because beneficiaries seeing LMHCs may also be seeing a psychiatrist, primary care physician, or other mental health provider. Instead, we intended to allow for within-group comparisons across time (pre- versus post-demonstration) for three reasons: first, to determine if there was a shift toward use of LMHCs; second, to determine how the demonstration impacted utilization among LMHC users; and third, to determine how the demonstration may have affected utilization among those seeing only non-LMHC MH provider types.

Definition of Measures. Using the variables available in the administrative claim records provided by TMA, we constructed several measures of interest: outpatient visit counts, inpatient episodes, expenditures for outpatient visits and inpatient episodes, and payments to providers. Our operational definition of each of these measures is in Appendix B.

Statistical Tests. All analyses were conducted using Statistical Analysis Software (SAS) version 8.02. To measure differences pre- and post- demonstration, where appropriate to the variable we used chi-square tests and tested differences in means with t-tests. To control for population differences, we used propensity score weighting to adjust the non-demonstration group population for differences in age, sex, member category, and interactions between these characteristics. Using the propensity score weights to control for variation in the only personal information we had available about the populations of interest, we compared weighted means across the two groups to test for statistical differences between the demonstration and non-demonstration areas on variables of interest. We first compared utilization across the two eligible populations, including the rate of any mental health care use and of counselor use. We then compared rates of use among those seeing a LMHC. To determine if the demonstration had a significant impact on the variables of interest, we used a difference-in-difference approach to determine whether the differences (e.g., in utilization or costs) between pre-demonstration and post-demonstration in the demonstration areas are significantly different than the differences between pre-demonstration and post-demonstration in the non-demonstration areas.

Survey of Beneficiaries

To assess the extent to which the changing of reimbursement policies for LMHCs impacts the health and treatment of covered beneficiaries under the TRICARE program, we designed a cross-sectional self-report survey. This cross-sectional survey was administered approximately nine months after full implementation of the demonstration. Using administrative claims data for mental health service users, we drew a random stratified (by catchment area and provider group) sample of 1,200 beneficiaries who met our definition of a mental health user (e.g., all respondents were adult users of mental health services). Our final response rate was 46 percent using various prompts and re-mailings (but no financial or other incentive). This response rate is among the highest in the range (between 6 percent and 47 percent) obtained in field tests of the Experiences of Care and Health Outcomes (ECHO) (Daniels et al., forthcoming). (See Appendix C for details on survey fielding methods.) Data collected allowed for a comparison of treatment outcomes of covered beneficiaries who receive mental health services from licensed or certified professional mental health counselors acting under physician referral and supervision, other non-physician mental health providers recognized under CHAMPUS and the TRICARE program, and physicians, with treatment outcomes under the demonstration project allowing independent practice of professional counselors on the same basis as other non-physician mental health providers.

Survey Content. An overview of the survey content is shown in Table 2.2. Much of the content was drawn from established and validated instruments used in both research and managed care. For example, we included key portions of the ECHO survey that was developed by the Consumer Assessment of Health Plans measurement team (Eisen et al., 1999; 2000). We also drew items from the Patient Health Questionnaire (PHQ) (Spitzer et al., 1999; Kroenke et al., 2001) to assess common mental disorders; the survey instruments used in the Partners In Care (PIC) (Wells et al., 2000) study for items about types of counseling or treatments received; and the Medical Outcomes Study (MOS) (Ware and Sherbourne, 1992) as a source for questions on attitudes about health care.

In addition, we also asked about some new and unique items to assess respondents' knowledge about the demonstration and exposure to the war in Iraq, which was ongoing during the field period, to understand their impact on mental health service use and outcomes. Because of the timing of the field period and the ongoing war, the evaluation of the demonstration effect is subject to confounding. In other words, it would be difficult to ascertain whether any effects we observe are due to the demonstration or to the war itself. Therefore, we thought it was critical to incorporate some measures of the war's impact into our

Table 2.2
Summary of Survey Content and Sources

Domain/Concept	Survey Item Source
Treatment for personal or emotional problems	ECHO
Counseling or treatment	PIC
Medication and other health remedies	PIC
Health plan and mental health benefits	ECHO
Health status	PHQ, ECHO
Attitudes about health and health care	MOS, PIC (DiMatteo et al., 1992; 1993; Link et al, 1991)
Knowledge of the TRICARE demonstration	New items developed for this study
Exposure to war in Iraq	New items developed for this study
Demographics	Standard demographic questions (gender, age, race, etc.)

evaluation by including a proxy for "war exposure." This would at the very least allow us to measure its impact and, where we observe variation, examine any demonstration effects over and above any differences due to the war.

Description of Measures. We derived a set of binary measures including those specified for Health Plan Employer Data and Information Set (HEDIS) indicators and four multi-item scales for selected item sets based on the literature. Scales included: (1) job stigma, (2) need for secrecy, (3) general adherence, (4) medication adherence, and (5) counseling adherence. Definitions and scoring rules for these variables are provided in Table D.1 of Appendix D.

Analysis. We created sample weights to adjust for differences across respondent age groups. To derive the weights, we first examined results from a logistic regression model that predicted response from a key set of variables we thought would affect findings (age group, provider type, gender, and demonstration region). In this model, only age group was a significant predictor of response/non-response. To adjust for this potential bias, we used the logistic regression model to predict the probability of response for all of the responders, and computed the non-response weight as 1/(predicted probability of response). All survey analyses are presented for the weighted data (e.g., with the sample size inflated to represent the distribution across age groups for the entire sampling frame).

Our first set of analyses examined the bivariate differences for beneficiaries who received mental health care services from a provider in the demonstration areas compared with those receiving services in the matched non-demonstration comparison areas. We used chi-square statistics to analyze differences for binary indicators and categorical measures, and we used t-tests to compare means for continuous measures. We then included key variables (e.g., indicator of demonstration status, demographics) along with clinical, service/treatment use, and attitude/perception variables in multivariable models if they were significant in the bivariate analysis. In addition to examining the impact of the demonstration, we also identified key factors associated with those outcomes. We also tested the impact of the Iraq war on TRICARE beneficiaries. We asked respondents whether any of their family members or close friends were deployed for the recent war in Iraq and, among those who said there were, whether any of the family members or friends were back from their tour of duty. These measures were included in multivariable analyses to evaluate the impact of war factors on service use above and beyond adjustment for other types of variation in the respondent sample. All analyses were weighted to reflect the survey sample of 1,200. Thus, our multivariable models adjusted for demographics, barriers to care, stigma, and impact of the Iraq war.

For these multivariable analyses, we selected a subset of outcome measures that we believed could have been affected by the demonstration. We included measures of access to mental health care (receipt of mental health care in the past six months, receipt of counseling from a mental health care provider in the past four weeks, taking any medication for a mental health problem in the past six months, and taking a prescription medication for a mental health problem in the past six months); adherence to mental health treatment (general adherence, adherence with medications, and adherence with counseling); indicators of mental health status (whether emotional or personal problems affected functioning, probability of having major depression, probability of having panic disorder, and probability of having somatic disorder); and selected HEDIS indicators of mental health care services (overall rating of counseling/treatment, whether respondents got urgent treatment as soon as needed, whether they got an appointment as soon as they wanted it, whether they got help by telephone, and whether they waited more than 15 minutes to see a clinician).

These binary indicators were scored from the ECHO items to assess consumer experience with specialty behavioral health care. Thus, the indicators have broader application because they identify current performance standards in managed behavioral healthcare organizations and are compatible with the National Commission on Quality Assurance (NCQA) accreditation requirements.

Qualitative Interviewing

We implemented a series of qualitative interviews with LMHCs and other relevant stakeholders regarding the implications and effects of independent LMHC practice under TRICARE. Our interviewing efforts were particularly designed to elicit data on five of the key issues posed by Congress in the NDAA.

- Administrative costs incurred as a result of required referrals to, and supervision of, LMHCs
- Effects of independent practice for LMHCs on confidentiality for TRICARE beneficiaries
- Effects of independent practice policies on LMHCs' willingness to participate as providers in TRICARE
- Any policy requests or recommendations regarding LMHCs made by health care plans or MCSCs participating in TRICARE.

Data Sources. To address the items listed above, we undertook three separate sets of interviews:

First, we spoke with TRICARE clinical providers, including LMHCs, clinical psychologists, and psychiatrists, from both the demonstration and non-demonstration regions. An initial round of baseline interviewing was undertaken with all of the providers at the beginning of the demonstration period. In addition, a follow-up round of interviewing was undertaken at the end of the project with those providers who participated in the demonstration. All of our interviews were semi-structured and based on formal interview protocols. (Copies of these protocols [baseline and follow-up] are available from the authors [Lisa_Meredith@rand.org or territ@rand.org] on request.) The focus of our interviews with clinical providers was on administrative costs related to practice requirements for LMHCs; on patterns of practice, supervision, and clinical outcomes in connection with practice requirements for LMHCs; and on patient confidentiality and communications practices as related to LMHC practice requirements.

Second, we undertook a separate set of interviews with TRICARE MCSCs responsible for administering mental health benefits. Again, we conducted baseline interviews with MCSC officials in both demonstration and non-demonstration regions and then did follow-up interviews with MCSC officials in the demonstration region. All of our MCSC interviews were semi-structured and based on formal interview protocols, and copies of these protocols also are available on request. The primary focus of these interviews was to investigate administrative costs to MCSCs associated with LMHC practice requirements, MCSCs' perceptions of effects on clinical outcomes and confidentiality associated with independent LMHC practice, and any related policy requests or recommendations made by the MCSC.

Last, we conducted several additional interviews with other stakeholders affected by TRICARE's practice requirements for LMHCs. In particular, we spoke with representatives from national counseling organizations (the ACA and the AMHCA), a representative from the Military Association of Officers Association of America (formerly known as the Retired Officers Association, a membership advocacy group), an official from the Clinical Quality Programs Division within the office of the Chief Medical Officer for TRICARE at Department of Defense, and congressional staff persons on defense oversight committees (with responsibility for the authorizing legislation for the TRICARE demonstration). These interviews were undertaken to obtain background information on practice by LMHCs, the historical origins of current administrative requirements in TRICARE, and potential policy implications for the TRICARE demonstration. These interviews were less structured and more open-ended than those involving clinical providers or MCSCs, because the purpose of these interviews was to provide context and background information, rather than primary data for evaluating results from the demonstration.

Analytic Approach. Qualitative data analysis for the evaluation was conducted primarily by generating matrices of interview findings and by examining responses to specific interview questions as aggregated by respondents' demonstration status (participating versus not participating) and by clinical profession (e.g., LMHCs versus other clinical providers). In addition, pre- and post-demonstration comparisons of interview findings were generated for those clinicians and MCSCs who actually participated in the demonstration. Based on the patterns of responses reflected in these matrices, we endeavored to address several major evaluation issues concerning the impact of the demonstration on administrative costs, confidentiality, willingness by LMHCs to serve as TRICARE providers, and related policy recommendations concerning LMHC practice requirements. In addition, where qualitative findings were relevant, we drew from those findings to supplement our interpretation of the quantitative data from our analyses of TRICARE claims and of survey responses of TRICARE beneficiaries.

Challenges Associated with the Evaluation

In late 2002, as DoD moved forward with efforts to implement this demonstration and we developed our evaluation strategy, the United States began major deployments in preparation for Operation Iraqi Freedom. At the same time, military personnel were still deployed in Afghanistan for Operation Enduring Freedom. Major combat operations in Iraq began in spring 2003, just as the expanded access demonstration was getting under way. Both the demonstration catchment areas and non-demonstration areas include military installations with deployable forces, both active duty and reserve components. While detailed data about the number of personnel deployed from these regions were not available to us, forces were deployed from both the non-demonstration and the demonstration areas during the course of this study.

As we outlined earlier, military life and related deployments can have a psychological impact on the families and loved ones of military personnel during peacetime as well as wartime. This psychological impact is likely to cause increased stress and could result in a higher need for mental health support and services. As a result, changes in mental health service utilization patterns among military health beneficiaries can be expected during major deployments and combat operations. It should, therefore, be recognized that the impact of the war in Iraq and the major deployments might confound any effort to isolate the impact the demonstration on utilization (and thus costs) of mental health care.

In an attempt to examine the potential impact of the war on mental health service need and utilization, we developed specific items for the survey of beneficiaries. We then aimed to use the data from those items in our multivariable models to examine differences in self-reported need, barriers to access, and service utilization between respondents from the demonstration areas and respondents from the non-demonstration areas.

Because the survey data could not be linked to the administrative claims data, and because there were no comparable administrative data available to us with respect to whether a particular beneficiary had a loved one deployed, we could not examine or control for the impact of the war in the administrative analyses of utilization and costs. Therefore, we offer caution here and again in presenting the results of our analysis that any increases in utilization and costs observed between the pre- and post-demonstration period in either the demonstration or non-demonstration areas could be a consequence associated with the war in Iraq and not just the demonstration.

It should be noted that the major deployments over the past several years might also impact the availability of mental health services for beneficiaries—e.g., if mental health personnel who were also reservists (and working in the civilian, purchased-care sector) were deployed, the number of available providers to treat military health system beneficiaries may decrease.

CHAPTER THREE

Impact on Beneficiaries

To address the evaluation objective of determining the extent of the demonstration's impact on outcomes, we developed and fielded a survey of TRICARE beneficiaries using mental health services in the demonstration and non-demonstration comparison catchment areas. In this chapter, we present data from the sample of 553 respondents who completed the survey. (A copy of the survey and details about its development and fielding procedures are in Appendix C.) To our knowledge, this is the first survey that has examined the perspectives of TRICARE beneficiaries who use mental health services. In addition, these data represent the only independent study to examine mental health symptoms and other factors related to use of mental health services for this population. This chapter also discusses the potential impact of the demonstration on beneficiary confidentiality.

Creation of Derived Variables

From the raw survey items, we created a set of derived variables that were used in the final analyses. (See Table D.1 in Appendix D for more information on these variables.). We include the scoring rules and show descriptive data for the overall sample of respondents—e.g., the mean and standard deviation for continuous measures or the percentage of respondents for binary measures. These variables include characteristics of the study design (e.g., an indicator of demonstration versus non-demonstration area, sample selection criteria, exposure to the demonstration); demographic characteristics (e.g., age group, gender, education, race/ethnicity), health characteristics (e.g., clinical status, functioning); use of mental health services and treatments (e.g., reported utilization, use of psychotropic medications); and perceived ability to access mental health care (e.g., perceived and experienced barriers to care, adherence to treatment, HEDIS indicators from the ECHO survey). We also included a question about personal experience with the recent deployment of a close friend or family member and a question about the extent to which this had an impact on the use of mental health services.

Data Analysis

All of the data presented in this chapter for the 533 respondents are weighted to represent the eligible sample of 1,200 beneficiaries. We present the weighted bivariate means (for continuous measures) or percentage (for binary indicators) comparing TRICARE beneficiaries in the demonstration catchment areas with beneficiaries in the non-demonstration catchment areas. Statistical significance for these two-group comparisons is shown in the form of t-tests for continuous measures or chi-square statistics for categorical or binary measures. We also present results from a set of multivariable regression models (ordinary least squares for continuous outcomes and logistic regression for binary outcomes). These models adjust for key design and demographic variables, variables that differed significantly by demonstration status, and other factors (e.g., barriers to mental health care, impact of Iraq war) that would be expected to affect outcomes. We highlight many of these findings in this chapter, and also summarize our analyses in Appendix D.

Survey Respondent Sample Description

Overall, the sample was evenly distributed across age categories (14 to 23 percent of the sample per age category), was predominantly female (82 percent), close to one-third had a college education (27 percent), and 81 percent were white. The majority of the respondents were U.S. born (89 percent) and a majority had children (80 percent). Of those with children, 24 percent reported that their children had also received counseling in the past six months. Only 12 percent lived alone, and about half (44.9 percent) were currently working. Surprisingly, a fifth of the respondents (20 percent) were not currently working due to health problems.

Comparison of Demonstration Versus Non-Demonstration Mental Health Services Users

Using responses from the survey of beneficiaries, we examined differences between the characteristics of mental health users in the demonstration catchment areas and those in the non-demonstration areas. In bivariate analyses (see Table 3.1 and Tables D.2 through D.13 in Appendix D), we found differences in several demographic characteristics by demonstration versus non-demonstration area. Beneficiaries in the demonstration areas were younger (χ^2 = 29.5, p < .001), more likely to be college educated (χ^2 = 4.2, p < .05), less likely to be African-American (χ^2 = 7.0, p < .01) and more likely to be white (χ^2 = 4.3, p < .05), less likely to live alone (χ^2 = 5.9, p < .05), and more likely to be currently working (χ^2 = 6.6, p < .05) compared with beneficiaries in the non-demonstration areas. Table 3.1 shows the demographic characteristics of survey respondents by demonstration area.

Figure 3.1 shows the extent to which mental health (MH) service users received particular types of care for their personal or emotional problems during the past six months. While the sampling frame was defined based on recorded use of mental health services, only 85 percent of the survey respondents reported having used some type of mental health service or treatment during the evaluation period. Most reported using some type of medication during this period (75.5 percent), and the same proportion reported taking a prescription (Rx) medication for their mental health problem (76.7 percent). Slightly more than half of the survey respondents (50.8 percent) reported having received counseling from a mental health specialist in the past four weeks. Very few of the beneficiaries in this survey reported using available alternative over-the-counter remedies (e.g., *Hypericum* [Saint-John's-Wort] [1.9 percent]).

Table 3.1
Demographic Characteristics of Survey Respondents

Characteristic	Non-Demonstration (%) (N = 282)	Demonstration (%) (N = 271)	t or χ^2
Age Group			29.46***
18–24	13.1	18.8	7.00**
25–34	19.5	18.8	0.08
35–44	19.1	23.5	3.38
45–54	18.8	21.4	1.24
55–64	16.5	11.3	6.59*
65+	13.0	6.2	15.67***
Male	19.1	17.0	0.94
Education			17.74**
High school or less	24.3	25.5	0.24
Some college	50.9	44.9	4.21*
College graduate	24.8	29.6	3.34
Latino ethnicity	6.0	6.0	0.00
Race			13.11*
White	82.4	86.9	4.33*
Black	10.9	6.5	7.04**
Other	6.7	6.6	0.00
U.S.-born	89.7	89.2	0.10
Have children	79.2	81.0	0.58
Child(ren) received mental health care	32.0	30.8	0.17
Live alone	14.9	10.2	5.93*
Working	41.9	49.3	6.55*
Not working due to health reasons	31.7	27.2	2.00

NOTES: * p < .05, ** p < .01, ***p < .001. Percentages may not add to 100 due to rounding.

In terms of differences in the demonstration areas' reported use of services, we found that beneficiaries in the demonstration areas were 4.3 percent more likely to have received mental health care within the past six months (χ^2 = 4.3, p < .05). However, we also found that beneficiaries in the demonstration areas were 8.1 percent less likely to report having received counseling from a mental health care provider (χ^2 = 6.5, p < .05).

Figure 3.2 shows the percentage of beneficiaries reporting that they talked to or saw a provider for counseling or treatment in the past six months, by type of provider. Respondents could have seen multiple types of providers, so we allowed for overlap. These provider utilization rates (ordered by prevalence) are based on 85 percent of the survey respondents (15 percent did not answer the related question). The figure shows that psychiatrists were visited at the highest rate (51.1 percent of beneficiaries), followed by psychologists (36.3 percent), and mental health counselors (34.3 percent) at a roughly equivalent rate. Nearly a quarter of the respondents (22.5 percent) reported seeing a primary care provider. Other mental health providers (psychiatric nurses, chaplain/religious counselors, marriage/family counselors) and social workers were visited at the lowest rates (15.9 percent and 13.2 percent, respectively). We did not find differences by demonstration versus non-demonstration respondents for use of psychiatrists or social workers but did observe differences between the two groups for other types of providers. Beneficiaries in the demonstration areas, as compared with those in the non-demonstration areas, were significantly less likely to use psychologists (χ^2 = 9.3, p < .01, more likely to use a primary care provider (PCP) (χ^2 = 13.8,

Figure 3.1
Respondents' Use of Mental Health Services and Treatments in the Past Six Months

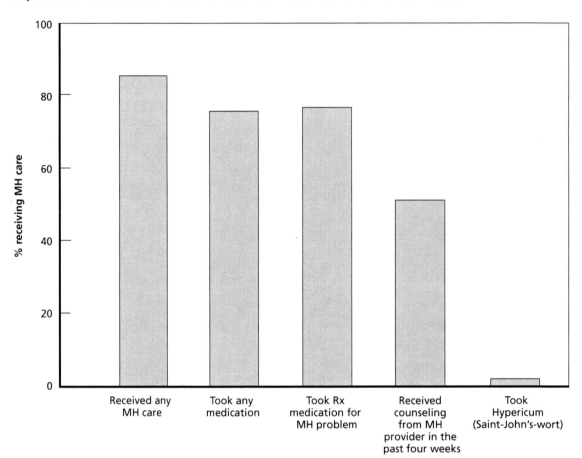

p < .001), and more likely to use other counselors (χ^2 = 5.1, p < .05), and there was a trend for slightly greater (not significant) use of LMHCs (χ^2 = 2.9, p < .10).

We also examined the distribution of provider types that respondents reported having seen most recently (not illustrated in the figure). These distributions are not directly comparable to the data in Figure 3.2 because the question about use in the past six months allowed for multiple responses, and the question about the most recent provider required only a single choice. However, the patterns are very similar, albeit recent use is less than use over the past six months, with the highest rates of use for psychiatrists, psychologists, and mental health counselors. There was a highly significant difference overall in this distribution by demonstration area (χ^2 = 30.4, p < .001), with the most striking differences for psychologists (less use in the demonstration areas) and LMHCs (more use in the demonstration areas).

We also found several differences in health and service use characteristics between respondents in demonstration and non-demonstration areas. For example, we found greater frequency in emotional or personal problems that affected functioning (72.6 percent versus 66.3 percent, χ^2 = 5.6, p< .01) among respondents in demonstration areas versus those in the

Figure 3.2
Type of Provider Respondents Saw for Counseling or Treatment in the Past Six Months

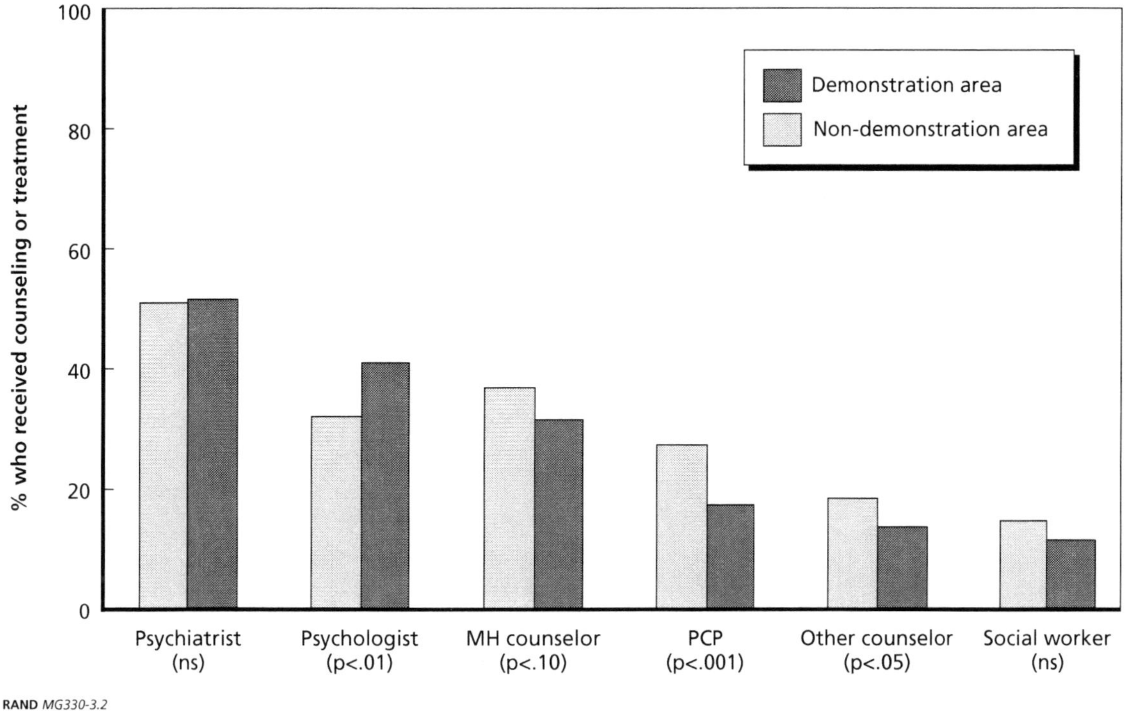

NOTE: ns = not statistically significant.

non-demonstration areas and more perceived barriers to mental care due to family-related problems (28.6 percent versus 19.4 percent, χ^2 = 13.2, p < .01). Beneficiaries in the demonstration areas also reported more use of mood stabilizers (9.1 percent versus 5.4 percent, p < .05) and antipsychotic medications (13 percent versus 6.0 percent, χ^2 = 4.7, p < .001) and lower use of benzodiazepenes (12.4 percent versus 18.2 percent, p < .01) than those in non-demonstration areas. In terms of HEDIS indicators of access to care, we observed a handful of differences. Beneficiaries in the demonstration areas were more likely to report improvement in dealing with daily problems (42.4 percent versus 36.6 percent, p < .05), getting urgent treatment as soon as it is needed (44.9 percent versus 28.5 percent, p < .01), and getting help by telephone (25.7 percent versus 28.5 percent, p < .01), but were less likely to report that they never waited more than 15 minutes for an appointment (55.7 percent versus 58.5 percent, p < .05). Among demonstration-area beneficiaries as compared with those in non-demonstration areas, there was a higher percentage with a close friends or family members deployed for the war in Iraq (34.5 percent versus 28.5 percent, χ^2 = 5.0, p < .05), and among those reporting deployments, a higher percentage reported that those who were deployed had not returned from duty (19.8 percent versus 14.4 percent, χ^2 = 6.2, p < .05). We found no bivariate differences by demonstration status in mental health symptoms or probable disorder, use of services and treatments, other barriers to care, or HEDIS indicators of access to mental health care.

To test the extent to which survey respondents who were TRICARE users of mental health services were aware of the changes made to expand access to LMHCs, we looked at their reported awareness. Overall, only 4.8 percent of beneficiaries knew about the demonstration project before receiving the survey, and while there was a slight trend for awareness to be higher among beneficiaries in the demonstration catchment areas compared with those in non-demonstration areas, this difference was not statistically significant (5.9 percent versus 3.7 percent, χ^2 = 3.2, p = .07).

Impact of Demonstration on Beneficiaries' Treatment Outcomes

In multivariable analyses, we observed little effect of the demonstration on beneficiary outcomes. We observed no differences by demonstration area in measures of access to mental health services (see Table D.14 in Appendix D), adherence to treatment (see Table D.15), or mental health status, including in the respondent's report of having symptoms of probable mental disorders such as depression, anxiety, panic, and suicidal ideation (see Table D.16). There were two exceptions to the limited effect on outcomes. Beneficiaries in the demonstration areas had a 32-percent lower likelihood of having received counseling from a mental health provider in the past six months (odds ratio [OR] = 0.68, 95 percent confidence interval [CI]: .51, .90, p < .01). Beneficiaries living in the demonstration also had a 36 percent greater chance of having emotional problems affect their functioning (OR = 1.34, 95 percent CI: 1.00, 1.81, p < .05).

We found a number of effects of the demonstration on HEDIS indicators of mental health services (see Table D.17). Living in the demonstration area was associated with nearly twofold greater odds of favorably rating counseling and treatment as a 9 or 10 on a 0–10 scale (OR = 1.95, 95 percent CI: 1.40, 2.70, p < .001), a greater chance of reporting an ability to "usually or always" get urgent treatment as soon as needed (OR = 3.97, 95 percent CI: 1.76, 8.95, p < .001), 1.5-times greater odds of being able to "usually or always" get an appointment as soon as it is wanted (OR = 1.54, 95 percent CI: .96, 2.50, p = .08), a more than threefold greater chance of saying that help could be received by telephone (OR = 3.59, 95 percent CI: 1.59, 8.12, p < .001), and a 46-percent less chance of never having to wait 15 minutes or more to see a clinician (OR = 0.54, 95 percent CI: .34, .86, p< .05). It should be noted however, that these differences may have existed prior to the demonstration period, particularly given that the demonstration areas were known to have high mental health service utilization prior to the demonstration and were chosen based on this utilization and provider availability.

Perceived Access to Mental Health Care

Other factors associated with access to mental health care include age group, perceived barriers to care, perceived workplace stigma, and whether the beneficiary had a family member or close friend deployed to the war in Iraq. Older beneficiaries were more likely to receive counseling and to be taking medication for a mental health problem (see Table D.14). For example, both those age 35–44 and those age 55 or over were twice as likely as those under age 25 to have received counseling (OR = 2.04, 95 percent CI: 1.27, 3.28, p < .01), and those age 44–54 were more than twice as likely to be taking a prescription medication for a mental health problem (OR = 2.43, 95 percent CI: 1.33, 4.42, p < .01).

Despite all respondents having a claim record for mental health service use, survey respondents with a higher score on the job-stigma scale (see the description of measures in Chapter Two) were less likely to have reported receiving mental health care (OR = 0.81, 95 percent CI: .69, .94, p < .01), and those who perceived that the stigma from seeking mental health care was a barrier to care were nearly three times as likely to be taking a prescription medication for a mental health problem (OR = 2.84, 95 percent CI: 1.80, 4.47, p < .001).

Another significant factor associated with access was whether anyone close to the beneficiary was deployed to the Iraq War. Deployment of a friend or family member was associated with a higher likelihood of receiving counseling from a mental health provider (OR = 1.74, 95 percent CI: 1.26, 12.41, p < .001) and a lower likelihood of taking a prescription medication for a mental health problem (OR = 0.58, 95 percent CI: 0.40, .84, p < .01).

Adherence to Treatment

Very few of the factors we studied were linked with adherence to treatment. Relative to the youngest group of beneficiaries, older beneficiaries scored higher on the medication-adherence scale. For example, beneficiaries age 25 or over were eight to ten times more likely to adhere to their medication regimens than those under age 25. In addition, beneficiaries who perceived that not being able to get help was a barrier to care had lower general adherence to treatment.

Mental Health Status

Figure 3.3 shows the percentage of survey respondents who reported having mental health problems, by type of problem (either probable disorder or problems that interfered with functioning). Close to 69 percent reported having an emotional or personal problem that made it difficult for them to work, take care of "things at home," or get along with other people. More than 45 percent of the respondents reported symptoms on the PHQ that indicate a high probability of having panic disorder. The probability of survey respondents having one of the other mental health disorders listed on the survey ranged from 8 percent to 26 percent.

We found that age was a significant predictor of mental health status. Being age 45 to 54 was associated with a twofold greater odds of reporting that an emotional or personal problem affected functioning, a more than threefold greater risk of having probable major depression, and a twofold increase in a probable somatic disorder relative to other age groups. Being a college graduate was associated with a lower likelihood of having a probable disorder (major depression, panic, or somatic) as was being African-American. Beneficiaries who were currently employed were 42 percent less likely to have panic disorder (OR = 0.58, 95 percent CI: 0.43, .77, p < .001). Perceived barriers also affected beneficiaries' reported mental health status. Perceived barriers due to family problems were associated with more problems in functioning (OR = 1.99, 95 percent CI: 1.31, 3.01, p < .01) and a greater likelihood of having major depression (OR = 1.81, 95 percent CI: 1.19, 2.75, p < .01).

Figure 3.3
Percentage of Respondents with Mental Health Problems, by Type of Problem

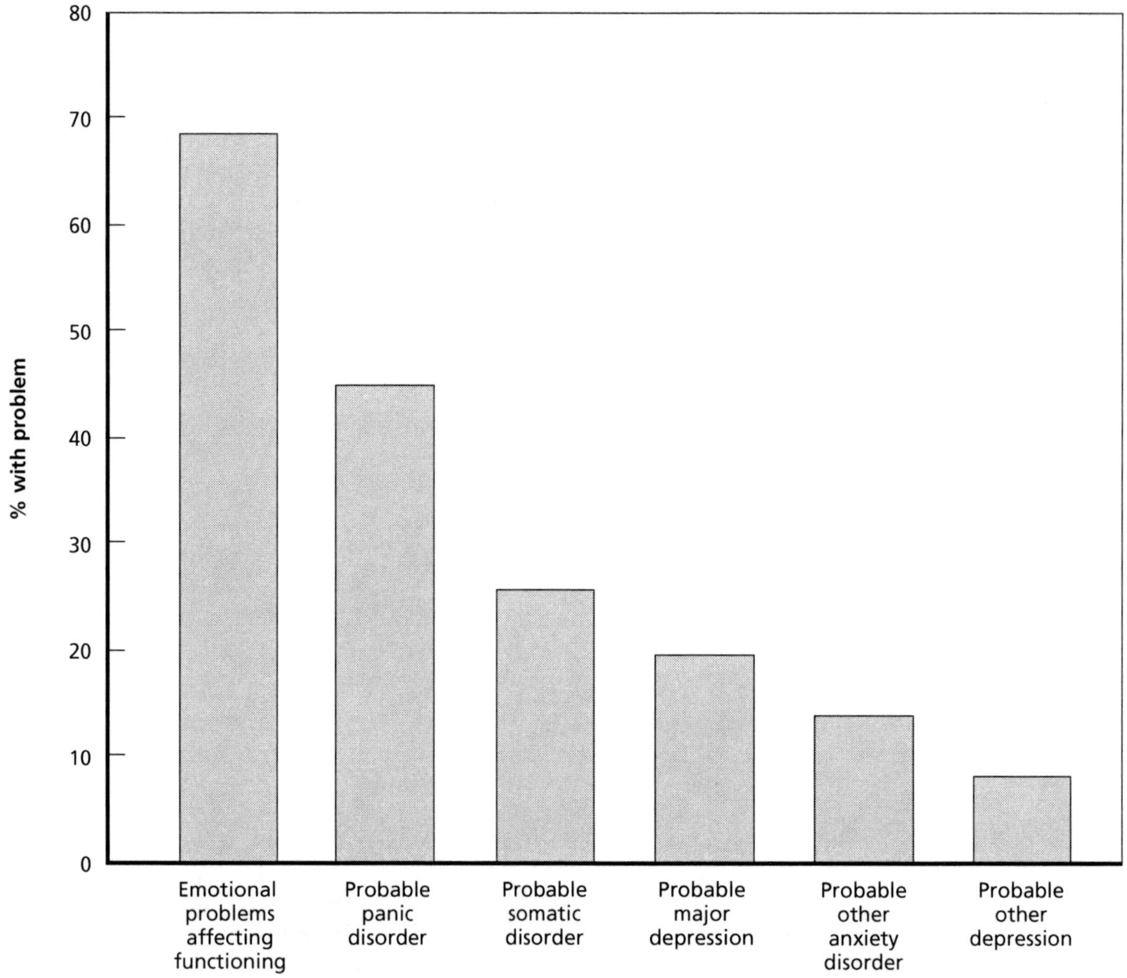

RAND *MG330-3.3*

Perception of an inability to find help was associated with a more than threefold greater odds of having major depression (OR = 3.43, 95 percent CI: 2.11, 5.58, p < .001) and a twofold odds of having a somatic disorder (OR = 2.04, 95 percent CI: 1.27, 3.25, p < .01).

Receiving mental health care due to the war in Iraq had a significant association with three of the four mental health status outcomes (emotional or personal problems affecting functioning, probable panic disorder, and probable somatic disorder, with the fourth being probable major depression) listed in Table D.16. Those who received mental health care for war-related reasons were five times more likely to have emotional or personal problems that affected functioning (OR = 5.01, 95 percent CI: 2.46, 10.17, p < .001), 3.89 times more likely to have probable major depression (p < .001), and 2.75 times more likely to have probable somatic disorder (p < .001).

Satisfaction with and Use of Mental Health Care Services

The overall weighted distribution of medication use is shown in Figure 3.4. Over half (52.7 percent) of the survey sample of mental health service users reported taking an antidepressant medication, whereas only 21.4 percent were taking some other non-mental health medication for a mental health problem. There was also a somewhat high rate (15.3 percent) of benzodiazapine (e.g., minor tranquilizers) use.

Additionally, beneficiaries who perceived barriers to access were significantly less likely to give their counseling and treatment high ratings (OR = 0.45, 95 percent CI: 0.30, .67, p < .001) and less likely to get an appointment as soon as they wanted one (OR = 0.26, 95 percent CI: 0.13, .50, p < .001), whereas beneficiaries who reported that professional circumstances were a barrier to care had more than three times greater odds of getting urgent mental health care as quickly as it was needed (OR = 3.27, 95 percent CI: 1.37, 7.82, p < .01, Table D.17).

Figure 3.4
Percentage of Respondents Taking Psychotropic Medications, by Type of Medication

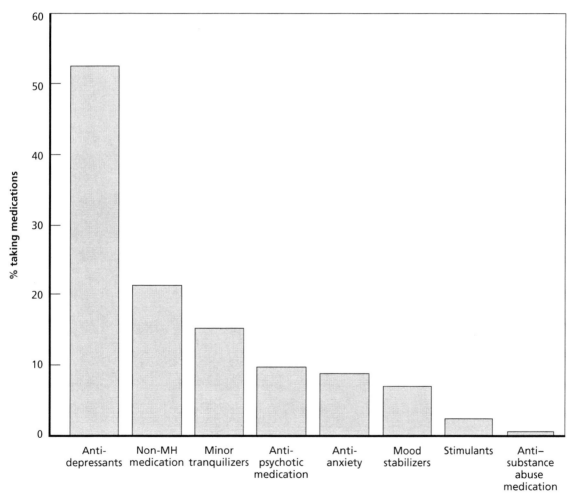

Impact of Iraq War

Across the entire survey sample, 31.5 percent reported that they had a close family member or friend deployed to the war in Iraq. Among those experiencing a deployment, 17.1 percent reported that the person had not yet returned from duty. Twelve-and-a-half percent of the survey respondents said that they had received mental health care due to the war.

We also ran a logistic regression model that predicted receipt of mental health counseling due to the war in Iraq to identify the factors associated with using mental health services for this reason. There was a slight tendency for survey respondents in the demonstration areas to make less use of mental health care due to the war in Iraq ($p < .05$) than respondents in non-demonstration areas, for older beneficiaries to be less likely to have received care for this reason ($p < .01$), for working beneficiaries to use less mental health care for this reason ($p < .001$), for perceived barriers due to cost to be associated with lower use of mental health care due to the war ($p < .05$), and for perceived barriers due to access or family-related problems to be associated with receiving less care. We also observed nearly 20 times greater odds of use of mental health services among those who had family members or close friends who were deployed (OR = 19.94, 95 percent CI: 11.22, 35.43, $p < .001$).

Impact on Beneficiary Confidentiality

The FY01 NDAA requested a description of the ways in which allowing for independent reimbursement of counselors affects the confidentiality of mental health and substance abuse services for covered beneficiaries under the TRICARE program. In this section, we summarize our findings on the potential impact of independent practice authority for LMHCs on beneficiary confidentiality.

LMHCs who provide clinical care to TRICARE beneficiaries are subject to the same legal privacy requirements as are all other health care providers under federal law. Pursuant to the Health Insurance Portability and Accountability Act (HIPAA) and the Privacy Rules of that legislation,[1] health care providers and health care plans have broad non-disclosure obligations in connection with personally identifiable health information. Providers (including LMHCs) are also required to take affirmative steps to protect the security of such information by implementing specified administrative, physical, and technical safeguards. The Privacy Rules include a number of exceptions that allow providers to disclose protected health information. Most important among them is an exception for "treatment, payment, and operations" (TPO), which permits clinical providers to use and share protected health information in the ordinary course of delivering healthcare.

In principle, one could imagine at least two potential effects on confidentiality as a result of independent practice by LMHCs. First, to the extent that clinical supervision is designed to ensure counselors' compliance with privacy requirements, removal of supervision might plausibly undermine that compliance. In practice, we found no evidence that the supervision requirement for LMHCs actually serves this purpose, nor did we find that the removal of supervision was associated with any change in confidentiality standards. Second,

---

[1] Health Insurance Portability and Accountability Act (HIPAA) of 1996, Pub. Law No. 104-191, 110 Stat. 1936 (codified as amended in scattered sections of 42 U.S.C.). Federal privacy and security rules enacted under HIPAA are codified at 45 C.F.R. Parts 160 and 164 (2004).

and at the other extreme, clinical supervision of LMHCs necessarily entails additional communications between providers and could involve additional record keeping by the supervisors and/or those being supervised. While any additional communication involving protected health information creates some incremental risk for wrongful or inadvertent disclosure, we found no direct evidence connected with the demonstration to show this kind of effect or in connection with supervised practice by LMHCs.

To investigate the effect on confidentiality of independent LMHC practice, we asked a series of related questions in our interviews with LMHCs, psychiatrists and psychologists, TRICARE MCSC executives, DoD officials, and representatives from professional organizations. In none of these interviews did we learn of any unique confidentiality issues or problems raised by the practice of LMHCs, whether supervised or independent. On a somewhat different note, a few providers did raise concerns regarding their uncertainty about what happens to patient information after it is communicated to TRICARE, and whether TRICARE has achieved compliance with all applicable HIPAA standards. These comments, however, were unrelated to the issue of independent practice by LMHCs.

CHAPTER FOUR

Impact on Providers

To understand the impact of the demonstration on TRICARE providers, we engaged in a series of interviews with TRICARE clinical providers in both demonstration and non-demonstration areas, both before and during the demonstration. We spoke with psychologists and psychiatrists as well as with LMHCs, and in those interviews we addressed topics ranging from the providers' perspectives on TRICARE referral and supervision requirements to the clinical roles of LMHCs in providing care to TRICARE beneficiaries.

The purpose of the interviews was to address several of the evaluation questions originally posed by Congress, particularly with regard to the impact of the TRICARE referral and supervision policies on LMHCs and the LMHCs' scope of practice. More specifically, the interviews explored the impact of TRICARE's policies in terms of their effects on administrative burdens and costs, providers' perceptions of autonomy, and quality of care provided to beneficiaries.

The dominant theme that emerged from the interviews was that the administrative requirement for physician referral was perceived as being particularly burdensome, and that the removal of that requirement made it easier for LMHCs to see TRICARE beneficiaries. Far less clear from the interviews, however, were any specific or actual administrative (financial) costs to LMHCs connected with the referral and supervision requirements, other than the use of their time. Several LMHCs described the administrative demands under TRICARE as being comparable to, or even less of a burden, than those under many private-sector health plans. On a different note, LMHCs described a broad range of baseline practices with regard to supervision under TRICARE, with some of them having engaged in very intensive supervision arrangements, and others describing much more sporadic or superficial experiences with supervision.

Interview findings generally suggested that major changes in the nature of care provided, or in the clinical roles of LMHCs, were not likely to result from the removal of referral and supervision requirements. Taken collectively, these findings suggest that the demonstration may have yielded modest administrative savings for some LMHCs under TRICARE, while leaving unchanged the scope and patterns of the LMHCs' practice, commitments to confidentiality, and other aspects of their practice.

Perceptions of Autonomy Among Counselors

Administrative Burden Associated with Referral and Supervision

We began our interviews with LMHCs by asking them to describe the referral and supervision requirements under TRICARE and their own administrative costs in complying with those requirements. Counselors from both demonstration and non-demonstration areas indicated that the baseline policy under TRICARE required patients to receive a referral from a physician, such as a psychiatrist or other physician, as a predicate to their being seen by an LMHC. LMHCs reported that, after obtaining that referral, their TRICARE patients were generally entitled to eight therapy sessions, with opportunity for more sessions based on a subsequent written authorization request made by the LMHC to TRICARE. Several of the counselors (from both demonstration and non-demonstration areas) said that the requirement that beneficiaries obtain a physician-referral in order to seek therapy from LMHCs had been a significant burden to their patients and an impediment to beneficiaries receiving care from LMHCs as opposed to other sorts of therapists (e.g., social workers, psychologists). Generally, though, this impediment was described as a discriminatory policy that made it harder for patients to access LMHCs, rather than as a source of administrative burden to LMHCs per se. Prior to the demonstration, none of the counselors identified the physician-referral requirement in itself as posing a substantial administrative burden or as generating costs directly to them. After the demonstration, LMHCs who participated did say that the demonstration had reduced the amount of time they previously spent in telephoning physicians to try to obtain, or to confirm, referrals to authorize therapy.

With regard to fulfilling TRICARE's baseline requirements for supervision, the LMHCs with whom we spoke described a range of supervision practices. Some indicated that they received regular supervision from physician or psychologist colleagues (particularly in mixed group-practice settings), while others indicated that supervision was minimal, not required of them, or (typically) limited to a review of session notes by a supervisor. Notably, two of the non-demonstration LMHCs with whom we spoke said that they did not believe they were required to receive supervision under TRICARE, and one said that were she required to spend time receiving supervision she would find it more difficult to afford to see TRICARE patients. For those LMHCS who participated in the demonstration, removal of the supervision requirement was reportedly not associated with major changes in their practice patterns or administrative burden or overhead. To the extent that LMHCs felt that they experienced administrative savings in the course of the demonstration, they tended to attribute those savings more to the elimination of the physician referral requirement rather than to the elimination of supervision. The theme that emerged from the interviews on supervision was that baseline supervision practices under TRICARE are highly varied, that some counselors are deeply committed to obtaining supervision regardless of TRICARE's requirements, and that in other instances compliance with the supervision requirement was more of a formality than a valuable exercise. In consequence, perhaps it should not be surprising that removal of the supervision requirement during the demonstration was not perceived as having a major effect by participating LMHCs. During our interviews with LMHCs, other mental health professionals, and managed care representatives, several respondents suggested that credentialing and licensing standards might be a more useful quality-control mechanism than the current TRICARE requirements for supervision and referrals.

To try to understand the administrative burden associated with LMHCs' baseline practice under TRICARE, we asked counselors some broad questions about their administrative practices and activities, and about their experience with the comparative administrative burdens of TRICARE and other private-sector insurers. The LMHCs described their administrative activities as generally involving the writing of session notes, the formulation of treatment plans, the filing of claims for payment, periodic communications with psychiatrists and other collaborators in treatment (including, presumably, supervision-related communications), and requests for authorization to TRICARE for more therapy sessions beyond the original set of eight preapproved sessions. Most of these types of activities were reportedly unaffected by counselors' actual experiences in the demonstration. Interestingly, more than half of the LMHCs', including both of those who actually participated in the demonstration, described TRICARE as being relatively easy to work with and non-burdensome from an administrative standpoint as compared with other insurers. Only one of the four counselors we interviewed expressed the opposite opinion.

We also asked LMHCs to try to estimate the amount of time that they spent each week on TRICARE administrative activities, and for those who participated in the demonstration, the amount of time that they ultimately felt was saved as a result of the provisional independent practice authority. These estimates proved to be difficult for counselors to formulate in a consistent way, because some of them carried very small TRICARE caseloads, some described receiving significant support from clerical assistants, and others drew a distinction between time spent on "ordinary" administrative activities versus appeals of disputed TRICARE claims. Notwithstanding these potential confounds in the analysis, for the four counselors who sought to answer this question, the average amount of time they reportedly spent on TRICARE administrative matters was about 10 to 15 minutes per patient per week. Both counselors who participated in the demonstration indicated that during the course of the demonstration they saved administrative costs by reducing the time spent seeking authorizations from physicians on behalf of TRICARE beneficiaries. One counselor estimated saving about one hour of related administrative time per week given a caseload of about 25 or 30 TRICARE patients who were seen weekly. The other counselor estimated saving about one hour of administrative time per TRICARE case over the lifetime of the case (the length of which was not specified). Both participating LMHCs described these administrative savings as making their practices under TRICARE significantly less burdensome than they had been prior to the demonstration.

Perceptions of Role Changes Among Counselors

In addition to asking LMHCs about the administrative costs and burdens of working with TRICARE patients, we also asked them several questions about the nature of their clinical practice, LMHCs' roles under TRICARE, and any likely advantages, disadvantages, or changes that they might anticipate as a result of eliminating the referral and supervision requirements. In general, the counselors described providing a broad range of psychotherapy services to adult, adolescent, and child clients. The majority of the LMHCs with whom we spoke did not feel that LMHCs needed to be supervised for these types of clinical activities, and several asserted that there was no reason for discriminating between LMHCs and other sorts of clinicians (e.g., social workers) on a professional basis.

The LMHCs uniformly expressed the opinion that there would be little change in their professional roles as a result of the removal of TRICARE referral and supervision requirements. Several noted that it would probably be easier and/or quicker for LMHCs to see TRICARE patients under the demonstration, and one of them suggested that public and professional perceptions about LMHCs might improve as a result of independent practice authority. None of the counselors identified any unique disadvantages accruing to unsupervised practice by LMHCs, but some did suggest advantages for TRICARE beneficiaries, including: (1) the possibility of more rapid access to crisis services, and (2) improved access to therapists generally during wartime mobilizations (when many TRICARE psychologists and psychiatrists might themselves be deployed overseas).

The two participating LMHCs with whom we spoke following the demonstration indicated that there had been no demonstration-related changes in their professional roles and activities, apart from reducing the administrative time they spent seeking physician referrals. Both perceived that the main effect of the demonstration had been to facilitate access by TRICARE beneficiaries, allowing beneficiaries to enter treatment more easily and more quickly. Based on their experiences under the demonstration, both participating counselors expressed the hope that TRICARE would remove the referral and supervision requirements on a permanent basis.

Perspectives of Psychologists and Psychiatrists on Independent Practice by LMHCs

To supplement our information on the potential administrative savings and clinical implications of independent practice for LMHCs, we also undertook interviews with several psychologists and psychiatrists practicing under TRICARE. We spoke with these providers for several reasons. First, we wanted to obtain some sense of the administrative activities and burdens of TRICARE practice, as perceived by mental health clinicians other than LMHCs. Second, we wanted to explore administrative issues relating to the supervision of LMHCs with some of the people who might actually perform a supervisory function (we should note that no formal documentation was readily available to indicate which providers actually conduct supervision of LMHCs, because there is no official paper trail of referrals or supervision). Third, we wanted to obtain some general impressions about LMHCs' practice and clinical roles from the perspective of those in allied professional disciplines.[1]

The psychologists and psychiatrists with whom we spoke had diverging opinions about the administrative burden of practicing under TRICARE. One psychologist and two psychiatrists described the administrative burdens associated with practice under TRICARE as being not very great, or no greater than those of other health plans. A second psychologist indicated that TRICARE is very burdensome in the procedures it requires for requesting additional therapy sessions (beyond the initially preapproved eight sessions). On a related note, one of the psychiatrists said that TRICARE's documentation requirements concerning medication management have been greatly simplified in recent years and are now very limited. He suggested that practice under TRICARE was likely to be more administratively burdensome for non-physician psychotherapists. In general, the psychologists and psychiatrists described administrative activities and record keeping for their TRICARE patients being similar to those for their non-TRICARE patients, as did LMHCs. Again, these activities include the writing of intake evaluations and session notes, the formulation of treatment plans, the filing of claims for payment, periodic communications with collaborators in treatment, and (at least for psychologists) requests for authorization to TRICARE for more therapy sessions be-

---

[1] We had also initially intended to speak with primary care physicians (PCPs) under TRICARE who, among other things, potentially serve as a major referral pathway for patients to LMHCs. In practice, however, no PCPs would take the time to speak with us about the TRICARE demonstration and the associated roles and responsibilities of LMHCs. Our experience suggests that practice issues relating to LMHCs are likely a very minor concern from the perspective of TRICARE PCPs, most of whose time and energy are devoted to other clinical and administrative challenges.

yond the original set of eight sessions. And again, the providers had difficulty in quantifying their own administrative costs associated with these tasks. One of the psychologists estimated that he spent about 10 to 15 minutes per TRICARE patient per week on related administrative activities.

With regard to supervising LMHCs under TRICARE, only one psychiatrist from among our four respondents actually had direct experience in performing such supervision. He indicated that LMHCs under his supervision had submitted written documentation to him about the treatments that they provided, and that he had been required to report the appropriateness of such documentation to TRICARE. The psychiatrist described this supervisory process as very burdensome and as "jumping through hoops." He also indicated that the administrative costs of his supervisory time were borne by the LMHCs that he supervised, not by TRICARE. The psychiatrist concluded that from his perspective, this system of supervision was not effective as a quality-control device for LMHCs, and he did not identify any specific concerns or disadvantages related to the prospect of unsupervised practice by LMHCs under the demonstration. Both the psychiatrist and a psychologist (both of whom practiced within a demonstration area) indicated that they had some experience in making treatment referrals to LMHCs. Neither felt that such referrals posed any significant administrative burden or costs from their point of view.

With regard to the scope of LMHC practice, LMHCs' general qualifications, and the advantages and disadvantages of eliminating referral and supervisory requirements under TRICARE, the psychologists and psychiatrists had mixed views. One psychologist said that he had no familiarity with LMHCs, their credentialing requirements, or their qualifications for independent practice. The other respondents all indicated that at least some LMHCs were qualified to provide independent treatment for at least some types of patients or psychiatric conditions, subject to having appropriate training and expertise. One provider said that he would refer patients only to LMHCs whom he personally knew were experienced and qualified to provide services. Another indicated that he would not send patients with cognitive impairments to LMHCs. Although one provider noted that the current supervision and referral requirements for LMHCs are not effective in ensuring quality of care, another pointed out that the credentialing rules for counselors in his state were very lax, and that removing the supervision requirement would carry the disadvantage of removing whatever (putative) quality controls that supervision might offer. A second provider agreed that removal of LMHC supervision and referral requirements would do nothing to ensure or improve the quality of care. He did suggest that elimination of the referral requirement might help some TRICARE patients to gain access to therapy more quickly than they would otherwise.

Provider Willingness to Participate in TRICARE

Beyond the issues described above, the NDAA FY01 also called for a description of the effect of DoD policies on the willingness of licensed or certified professional mental health counselors to participate as health care providers in the TRICARE program. During our qualitative interviews with representatives from the ACA and AMHCA, the lack of independent practice authority for LMHCs was cited as a major reason why their members indicated an unwillingness to join the TRICARE provider networks. While these organizations had no quantitative data available to assess the effect of this particular DoD policy, the representatives noted that this issue was among the concerns most frequently cited by their members.

To evaluate the impact of the demonstration (which, as discussed earlier, offered independent practice authority for LMHCs) in encouraging LMHCs to participate in TRICARE, we reviewed the trends in the number of LMHCs participating in the demonstration as well as the trends in the number of LMHCs enrolled as networked TRICARE providers (see Table 4.1). [2]

We examined two sources of data from TriWest (the MCSC responsible for the TRICARE network in the demonstration areas). To obtain information on trends in the number of LMHCs participating in the demonstration, we relied on the monthly reports provided by TriWest to TMA. Beginning with its August 2003 monthly report, TriWest also began to indicate which of the demonstration participants were enrolled as network providers (that is, LMHCs who were enrolled as TRICARE preferred providers—i.e., providers who have agreed to take a negotiated lower rate for services). Therefore, in Table 4.1, we also present the percentage of demonstration participants who were TRICARE network–enrolled providers. As shown in the table, the number of demonstration participants increased during the first few months of the demonstration but then leveled out during the middle of the demonstration period, likely due to the fact that TMA used only one mailing to advertise the demonstration opportunity to LMHCs. During the demonstration period, the number of network-enrolled LMHCs steadily and modestly increased in both regions serving the demonstration catchment areas. Unfortunately, data on the number of enrolled LMHCs in the non-demonstration catchment areas were not made available and therefore cannot be used

Table 4.1
LMHC Participation in Demonstration and in TRICARE Network, by Demonstration Areas and Month

Month	Colorado Springs		Omaha	
	Number of LMHCs Who Were Demonstration Participants/(% of participants who also participate in network)	Number of LMHCs Enrolled in Network	Number of LMHCs Who Were Demonstration Participants/(% of participants who also participate in network)	Number of LMHCs Enrolled in Network
January 2003	41	99	41	88
February 2003	57	100	53	89
March 2003	62	101	55	90
April 2003	64	101	55	92
May 2003	67	101	55	92
June 2003	68	103	55	92
July 2003	68	104	55	92
August 2003	68 (59)	105	55 (53)	92
September 2003	67 (59)	107	55 (53)	92
October 2003	66 (59)	107	55 (53)	91
November 2003	66 (67)	108	55 (55)	91
December 2003	66 (67)	109	55 (55)	96

---

[2] Enrollment as a TRICARE network provider implies that the provider has agreed to serve as a preferred provider for TRICARE Extra beneficiaries and accept network reimbursement rates. It should be noted, however, that any LMHC who is authorized to provide services under TRICARE can provide services and receive reimbursement.

for comparison purposes. As such, we cannot examine the extent to which the temporary in-dependent practice authority may have influenced the modest increase in the number of network-enrolled LMHCs during the demonstration period. It is also important to note that whether or not providers are likely to enroll as TRICARE network providers is likely also a function of their willingness to accept the in-network reimbursement rate for their services rather than solely a function of practice authority.

CHAPTER FIVE

Impact on TRICARE

As discussed in Chapter Two, expanding access to mental health counselors might be expected to impact the TRICARE program in a number of ways. First, opening up access to mental health services might change the volume and type of users, as well as the volume of use and costs of mental health care provided to TRICARE beneficiaries. Second, changing administrative procedures for LMHCs might also have an impact on the administrative costs associated with the delivery of mental health care. This chapter provides data on the impact the demonstration had on the TRICARE program overall, in terms of utilization and costs of mental health care.

For comparison purposes, we present data on beneficiaries in demonstration and non-demonstration catchment areas. As mentioned earlier in the report, the demonstration catchment areas included Offutt Air Force Base (Nebraska), U.S. Air Force (USAF) Academy (Colorado), and Ft. Carson (Colorado); the non-demonstration catchment areas include Wright-Patterson Air Force Base (Ohio), Luke Air Force Base (Arizona), and Ft. Hood (Texas).[1] The pre-demonstration period is defined as the one-year period beginning January 1, 2002, and ending December 31, 2002. The post-demonstration period is defined as the period of the demonstration's implementation and includes the one-year period beginning January 1, 2003, and ending December 31, 2003. We use administrative data from TRICARE claims to describe the level and cost of mental health care use over this period. We then present a difference-in-difference analysis designed to assess the impact of the demonstration on utilization and costs of mental health care.

Table 5.1 provides a brief overview of the number of eligible beneficiaries and users of mental health services in the demonstration and non-demonstration areas during the years of study. As noted, in 2002, there were 12,462 unique mental health service users in the demonstration area and 19,965 in the non-demonstration areas. The number of individuals who met our inclusion criteria increased in both the demonstration and non-demonstration areas during the demonstration period (2003). As a percentage of eligible beneficiaries, mental health service users rose from 9.3 percent to 10.1 percent (χ^2 = 57.05, p < .0001), and non-demonstration users rose from 9.6 percent to 10.4 percent (χ^2 = 58.70, p < .0001), during the demonstration.

---

[1] See Appendix B for additional details on the selection of these catchment areas.

Table 5.1
Eligible Beneficiaries and Mental Health Services Users by Area and Year

	Demonstration Areas		Non-Demonstration Areas	
	Pre-Demonstration	Post-Demonstration	Pre-Demonstration	Post-Demonstration
Total number of eligible beneficiaries (18 years or older)[a]	134,616	137,187	208,770	212,794
Total number of mental health service users[b]	12,462	13,876	19,965	22,154
Users as a percentage of eligible beneficiaries	9.3%	10.1%	9.6%	10.3%

[a]Data on the actual number of eligible beneficiaries were drawn as of April 30 of the study year. The number of eligible beneficiaries can change throughout the year as new beneficiaries become eligible or ineligible for TRICARE coverage.

[b]*Mental health service user* is defined broadly to include anyone 18 years or older who, during the year, saw a mental health care provider, had a mental health diagnosis on at least one claim, received a mental health service, and/or filled a prescription for a psychotropic medication (see Chapter Two for a fuller description of this definition).

Demographic Characteristics of Users

Table 5.2 describes the demographic characteristics of the mental health service users by demonstration area and year of study. Data on race and marital status are not presented (NP) due to the very high frequency of "missing" data in the files provided by DoD.[2] (Table E.2 in Appendix E provides a breakdown of demographic characteristics by users and non-users in each year.)

In the following discussion, we refer to the tables in Appendix E, which provides in-depth data on mental health services users and non-users. Table E.2 provides a breakdown of demographic characteristics by users and non-users in each year. As compared with the non-demonstration areas, there was a higher percentage of beneficiaries in the demonstration areas who were active duty (AD), dependents of active duty (ADD) beneficiaries, or dependents of retirees (RDD), and fewer who are over 65 years of age. It should be noted that these differences exist in both the mental health service user and non-mental health service user beneficiary population and likely reflect the differences associated with these catchment areas. For example, the student population at the USAF Academy would likely influence the age distribution in the demonstration region that includes that catchment area. It should also be noted that compared with the whole eligible population across the groups, mental health users are more often female, dependents of active duty or retirees, and between the ages of 18 and 45 (also see Table E.2).

For purposes of the analyses presented in this chapter, we separated mental health service users into four analytic groups based on the type of providers from whom they received outpatient services. To isolate beneficiaries who received services from LMHCs for purposes of comparison and to eliminate overlap among groups, we grouped beneficiaries

---

[2] Rates of "missing" data on race and marital status did not differ between users and non-users, across demonstration and non-demonstration areas, or across pre- and post-demonstration.

Table 5.2
Demographic Characteristics of Mental Health Services Users by Area and Year

	Demonstration Area		Non-Demonstration Area	
	Pre-Demonstration Number of Users/(%)	Post-Demonstration Number of Users/(%)	Pre-Demonstration Number of Users/(%)	Post-Demonstration Number of Users/(%)
Gender				
Female	8,472 (68%)	9,453 (68.1%)	13,917 (69.7%)	15,469 (69.8%)
Race	NP	NP	NP	NP
Marital status	NP	NP	NP	NP
Member category/type				
Active duty	594 (4.8%)	585 (4.2%)	540 (2.7%)	573 (2.6%)
Active duty dependent	2,326 (18.7%)	2,663 (19.2%)	3,360 (16.8%)	3,695 (16.7%)
Retired	2,897 (23.2%)	3,274 (23.6%)	4,786 (24.0%)	5,387 (24.3%)
Retiree dependent	5,162 (41.4%)	5,727 (41.3%)	8,889 (44.5%)	9,891 (44.6%)
Student/other	235 (1.9%)	349 (2.6%)	316 (1.6%)	464 (2.1%)
Missing	1,248 (10.0%)	1,278 (9.2%)	2,074 (10.4%)	2,144 (9.7%)
Age				
18–24	1,598 (12.8%)	1,774 (12.8%)	2,089 (10.5%)	2,258 (10.2%)
25–34	1,467 (11.8%)	1,778 (12.8%)	2,228 (11.2%)	2,469 (11.1%)
35–44	1,948 (15.6%)	2,064 (14.9%)	2,508 (12.6%)	2,696 (12.2%)
45–54	2,108 (16.9%)	2,306 (16.6%)	2,972 (14.9%)	3,301 (14.9%)
55–64	1,724 (13.8%)	1,954 (14.1%)	3,020 (15.1%)	3,433 (15.5%)
65 and over	3617 (29.0%)	4,000 (28.8%)	7,148 (35.8%)	7,997 (36.1%)

NOTE: Percentages may not add to 100 due to rounding.

into only one category even if they received services from more than one provider type during the year. Using a hierarchical approach, we devised the following groups: by LMHC first, followed by psychiatrists, non-physician OMH providers, and finally by "other physicians" (e.g., primary care, internal medicine).[3]

We used this hierarchical approach to isolate those beneficiaries who received care from LMHCs as the primary group of interest and to then eliminate overlap among the groups; however, it should be noted that beneficiaries in the LMHC, OMH provider, and psychiatrist groups might have also received care from another type of mental health care provider. It should also be noted that the number of beneficiaries who saw other-physician providers are individuals who met our inclusion criteria based either on a claim for a psychotropic medication (we included only those medications routinely provided for psychotropic uses) or on having a mental health diagnosis listed on a physician claim but not having any claims for visits to a mental health service provider during the year of study.[4]

Table 5.3 shows how users were distributed across these hierarchical groups. As a proportion of mental health service users who met our inclusion criteria, those who saw

---

[3] The data to create these groups were drawn from the administrative claims submitted to TRICARE for care rendered in the purchased care system; that is, if the beneficiary saw a provider only inside the MTF, records associated with that visit were not in the claim files we used for these analyses.

[4] Individuals who met our inclusion criteria but who did not see a mental health provider (for example, they met our inclusion criteria based on having a mental health diagnosis on a claim during the year or received a psychotropic medication) were grouped in the "other physician" category. However, some of these individuals did not have a claim for a mental health–related outpatient physician visit.

Table 5.3
Mental Health Services Users by Type of Service Provider

	Demonstration Areas		Non-Demonstration Areas	
	Pre-Demonstration Number of Users/(%)	Post-Demonstration Number of Users/(%)	Pre-Demonstration Number of Users/(%)	Post-Demonstration Number of Users/(%)
Total number of users	12,462	13,876	19,965	22,154
Saw an LMHC[a]	603 (4.8%)	750 (5.4%)	595 (3.0%)	700 (3.1%)
Saw an OMH provider	2,050 (16.5%)	1,897 (13.7%)	1,959 (9.8%)	2,160 (9.7%)
Saw a physician				
Psychiatrist	1,527 (12.3%)	1,747 (12.6%)	2,815 (14.1%)	2,918 (13.2%)
Other physician	8,282 (66.5%)	9,482 (68.3%)	14,596 (73.1%)	16,376 (73.9%)

NOTE: Percentages may not add to 100 due to rounding.
[a]Includes pastoral counselors, although visits to pastoral counselors were extremely rare across the sites and years.

LMHCs represent 4.8 percent and 3.0 percent of users during the pre-demonstration period in the demonstration areas and non-demonstration areas, respectively. During the demonstration period, these proportions rose to 5.4 percent (χ^2 = 4.32, p = 0.04) and 3.1 percent (χ^2 = 1.14, p = 0.29), respectively. The percentage of users seeing a psychiatrist (but not an LMHC) rose, but not significantly, in the demonstration areas (from 12.3 percent to 12.6 percent; χ^2 = 0.68, p = 0.41) and fell significantly in the non-demonstration areas (from 14.1 percent to 13.2 percent; χ^2 = 7.70, p = 0.006). The percentage of users seeing a mental health services provider other than an LMHC or psychiatrist fell in both areas, with a significant change in the demonstration areas only (16.5 percent to 13.7 percent; χ^2 = 39.80, p <.0001). The percentage of users not seeing any mental health provider was significantly higher in the non-demonstration areas in both the pre-demonstration (73.1 percent versus 66.5 percent; χ^2 = 163.31, p < .0001) and post-demonstration (73.9 percent versus 68.3 percent; χ^2 = 131.35, p < .0001) periods, and increased in both groups of areas (demonstration areas—χ^2 = 10.52, p = .001; non-demonstration areas—χ^2 = 3.55, p = .06.) The percentage of users seeing each of the mental health provider types in the non-demonstration areas was correspondingly lower in both periods, with the exception of those user seeing a physician (psychiatrist or other physician) in the post-demonstration period (where the percentage seeing a psychiatrist or other physician in the non-demonstration areas in the post-demonstration period was higher than the percentage seeing a psychiatrist or other physician in the demonstration areas post-demonstration).

Using the same provider-based analytic groups, we broke down the demographic characteristics of mental health services users by year (see Table E.3 in Appendix E). The distribution of age and member category among mental health services users varied significantly by provider group across both years and areas, with those users seeing only non–mental health physicians (called "other physician" hereafter) more likely to be over 65, retired or retired dependents, and male than those seeing any of the mental health providers.

In Table E.4, we present the distribution of users by mental health diagnosis (diagnoses were reported on the administrative claims and are grouped according to diagnostic groups from the *Diagnostic and Statistical Manual,* Fourth Edition [DSM-IV], American Psychiatric Association, 1994). As shown in that table, the distribution of mental health diagnoses within the study year are significantly different (using χ^2 tests; p < .0001) across provider groups. For example, mood disorders are the most common of the mental health diagnoses among users who see psychiatrists and those who see LMHCs (e.g., 71.3 percent and 64.3 percent of demonstration-area mental health services users in the pre-demonstration period, respectively). Adjustment disorders are the most common diagnoses among those who see OMH providers (e.g., 48.0 percent in the demonstration areas and 56.7 percent in the non-demonstration areas at pre-demonstration). These patterns held across demonstration and non-demonstration areas both pre- and post-demonstration.

Description of Utilization

One of the questions to be answered under the FY01 NDAA legislation was what effect, if any, the demonstration had on utilization of mental health services provided by LMHCs, OMH providers, and physicians. In this section, we provide estimates of utilization of mental health care within each of the analytic groups of interest. Again, these data are based on administrative claims paid by TRICARE for services rendered in the purchased-care sector during the years of study. In Table E.5, we provide data on the type of care provided to these mental health services users by provider group in each study year. We provide data on the overall volume of visits, per year and per month of study, for both outpatient and inpatient use for users in each provider group, and the mean number of visits per month and per year, in Tables E.6 and E.7.

Visits for Mental Health Services

The overall volume of mental health–related visits for mental health services users by provider group, year, and area are shown in Table E.6 (for a definition of how we defined and counted these visits, see Appendix B). In the post-demonstration year, the overall number of unique beneficiaries seen and volume of outpatient visits per year increased in both the demonstration and non-demonstration areas for every provider group except those in the OMH provider group within the demonstration areas, where the number of unique mental health services users decreased from 2.050 to 1.897. Figure 5.1 displays the mean number of mental health visits per year by users in each provider group. As noted, the mean number of mental health visits by people seeing LMHCs decreased during the demonstration period in the demonstration areas and the non-demonstration areas, although the change in either group was not statistically significant. The average number of mental health visits remained the same or increased slightly during the demonstration period for all other provider groups, with the only significant increase in the other-physician group in the non-demonstration areas (t = 3.91, p = 0.0001).

Type of Outpatient Mental Health Care Provided

We also examined the types of mental health care provided to users in each provider group by area and by year. Table E.5 provides a description of the characteristics of the treatments provided to users, including whether they received psychotherapy alone, psychotherapy in combination with medication, or medication alone. We also present the distribution of users who filled a prescription for a psychotropic medication and the mean number of psychotropics per year for users in each provider group, area, and year.

Figure 5.1
Mean Number of Mental Health Visits per Year by Mental Health Services Users

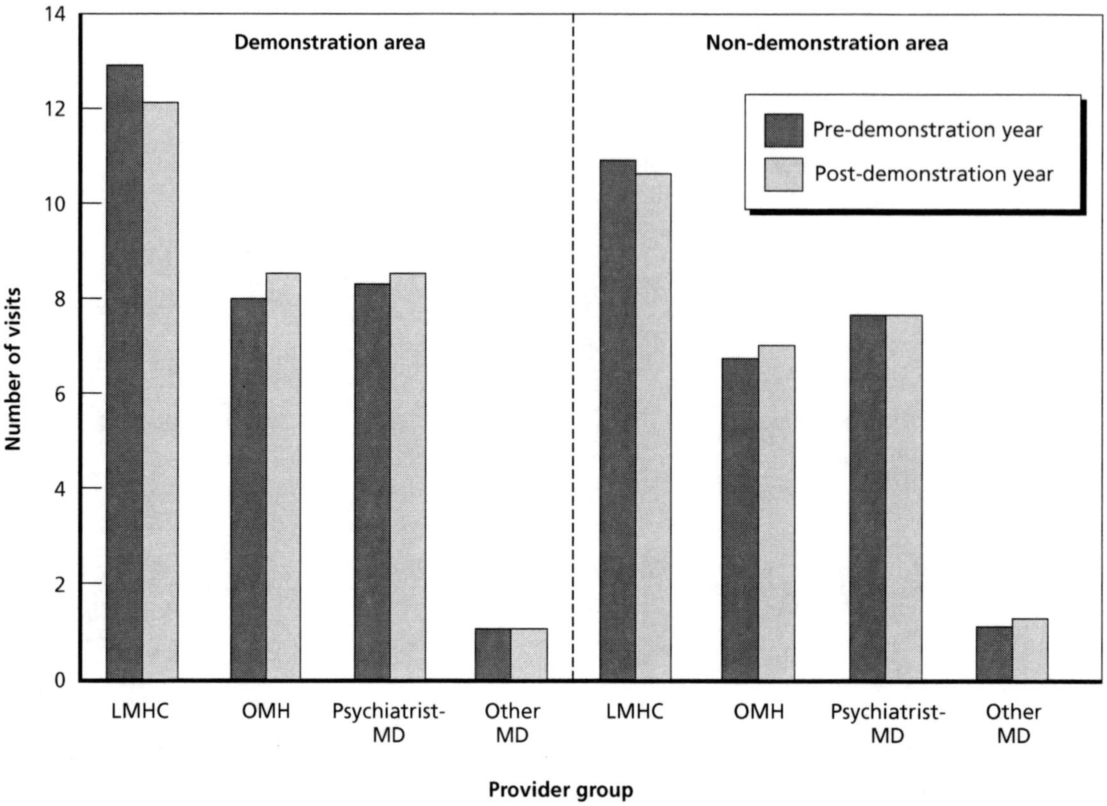

RAND MG330-5.1

The mean number of psychotropics per year for users who saw LMHCs in the demonstration areas decreased from 2.01 to 1.53 (t = 4.71, p < .0001), with the percentage taking any psychotropic drug falling from 73.3 to 65.2 (t = 4.22, p < .0001). There was no corresponding significant decrease in any of the other provider-type groups or in the non-demonstration group, suggesting that the decrease may be due to the removal of the requirement that LMHCs have oversight by a physician (who could prescribe a psychotropic drug). (See "Effects of the Demonstration" below for a difference-in-difference analysis of the significance of this outcome.) The most common types of medication taken by users in these areas were antidepressants (the percentage of users taking antidepressants ranged from 75 to 95 depending on the area and provider group), followed by benzodiazepines (ranging from 35 percent to 45 percent). Use of antipsychotic medications was more common among users who saw psychiatrists (28.3 percent in the demonstration areas and 21.3 percent in the non-demonstration areas in the pre-demonstration period) than among those mental health service users who are in the other provider groups.

Inpatient Mental Health Care Among Outpatient Mental Health Services Users

While our sample of mental health services users is grouped according to use of provider type for mental health care in an outpatient setting, we also examined the pattern of inpatient mental health care (for an explanation of how we defined and counted inpatient episodes, see Appendix B). Table E.5 provides a description of the number of users who received inpatient mental health services, the mean number of episodes per user per year, and the mean length of stay for these inpatient episodes per user per year.

In the pre-demonstration period, beneficiaries who saw LMHCs had an average of 0.13 inpatient episodes per user per year. This number decreased slightly in the post-demonstration period to 0.11 inpatient episodes per user per year (t = 0.84, p = 0.40), while beneficiaries who saw LMHCs in the non-demonstration areas during the same time frame had a slight non-significant increase from 0.13 to 0.17 (t = 1.61, p = 0.11) in the mean number of inpatient episodes per user per year. In the demonstration areas, the mean number of episodes increased significantly from 0.06 to 0.09 visits per user per year for the OMH provider group (t = 2.20, p = 0.03) and from 0.13 to 0.18 visits per user per year for the psychiatrist group (t = 2.10, p = 0.04). Changes in the other-physician provider group and in the groups in the non-demonstration areas were not statistically significant.

The mean length of stay for inpatient care users in the LMHC group increased in both the demonstration and non-demonstration areas; however, the changes were not statistically significant. For these groups, the mean length of stay rose in the demonstration areas (from 5.68 days per user per inpatient stay to 6.68 days per user per inpatient stay; t = 0.83, p = 0.41) and in the non-demonstration areas (from 5.16 days per user per stay to 5.58 days per user per stay; t = 0.34, p = 0.74). The only significant change in the mean length of stay was an increase from 7.6 to 9.8 days among the other-physician provider group in the non-demonstration areas (t = 3.90, p < .0001).

Overall Health Care Use by Mental Health Services Users

Overall health care use by mental health service users (outpatient visits and inpatient admissions for mental health and non–mental health care together) also increased in both the demonstration and non-demonstration areas for every provider group (see Table E.6). Figure 5.2 shows the mean number of outpatient visits made by users for any health care service by area and provider type. The mean number of hospital admissions per user per year is shown in Table 5.6. There were statistically significant increases in mean visits by users seeing OMH providers (t = 2.87, p = 0.004) and users seeing psychiatrists (t = 2.09, p = 0.04) in the demonstration areas and by users seeing other-physician providers (t = 2.74, p = 0.006) in the non-demonstration areas.

Description of Expenditures

As utilization changes, so can the costs associated with rendered care. As more care is consumed, the overall expenditures for mental health services also rise. To examine the impact of the demonstration on expenditures for mental health care, we examined overall expenditures by the government for outpatient mental health visits and inpatient mental health episodes as well as expenditures for all health care (mental health and non-mental health) paid by TRICARE for users by area and year of study (see Table E.8).[5] We also provide data on the total and average payments made to providers by the government for care rendered to users during the years of study (see Table E.9).

---

[5] Expenditures were not adjusted for inflation because no significant differences were observed between the years of study.

Figure 5.2
Mean Number of General Health Care Outpatient Visits by Mental Health Services Users, Pre- and Post-Demonstration Years

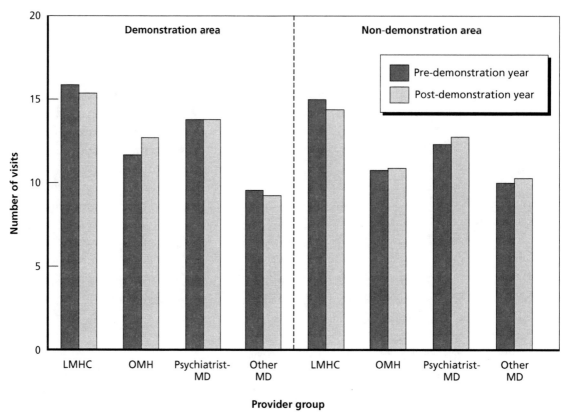

RAND *MG330-5.2*

Expenditures for Mental Health Care

As expected, given the increases in the number of beneficiaries who sought mental health care in post-demonstration period (as compared to the number in the pre-demonstration period) in both the demonstration and non-demonstration areas, there was an increase in the overall total expenditures related to mental health care (outpatient and inpatient) for mental health users within each provider group. Mean expenditures on MH care per user also increased for all provider groups in the demonstration and non-demonstration groups, with one exception. For those mental health users in the LMHC group in the demonstration areas, the mean expenditure for outpatient MH care visits per user decreased nonsignificantly from $802 per user per year to $749 per user per year (t = 0.81, p = 0.42) in the post-demonstration period (see Figure 5.3 and Table E.5). The increase in mean costs in the OMH provider group was statistically significant at the 95 percent confidence level or greater in both areas, as was the increase in the other-physician group in the non-demonstration area.

Figure 5.3
Mean Expenditures per Mental Health Services User for Outpatient Mental Health Care

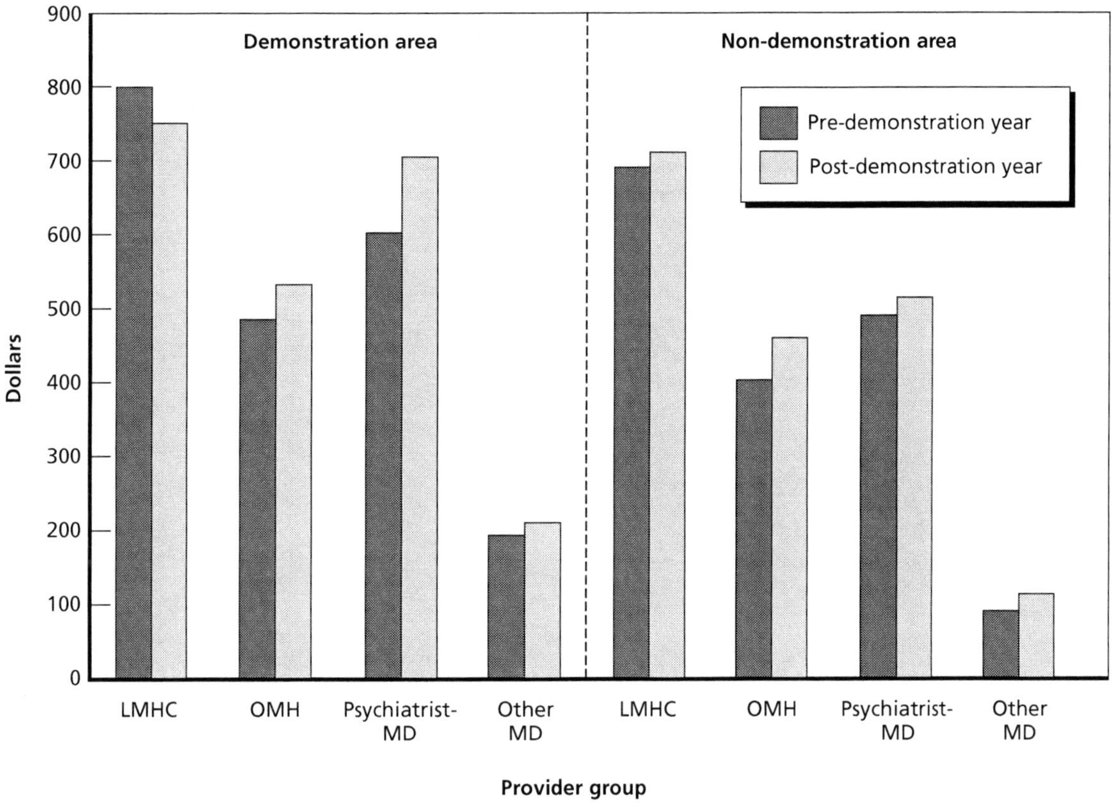

RAND MG330-5.3

Similarly, overall total expenditures for all health care (outpatient and inpatient, mental health and non-mental health) received by users in both demonstration and non-demonstration areas within all provider groups increased as overall health care use increased (utilization patterns are reported in Table E.5).

Payments to Providers

We also sought to determine whether the payments made to each provider group were affected by the demonstration. To do so, we examined the payments for visits made by users to each provider group by area and year. Unlike the analysis above, in which we summarized visits and payments by a hierarchical grouping of providers that each patient saw over the course of the year, we grouped visits and payments by provider type for services provided to beneficiaries who saw each of the mental health care provider types (see Table E.9). We distributed each user's visits and costs across Table E.9 into the columns corresponding to the types of providers from whom the user received care. As seen in the table, the overall number of visits to each provider group increased in each area and year, resulting in an increase in the overall total payments made to these provider groups. In the demonstration areas, changes in mean visits and payments to most provider types were not significant at the 95-percent confidence level. The only exception was for payments to other-physician providers, which increased from \$168 to \$198 per year per user (t = 2.18 p = 0.03). In the non-demonstration areas, visits to psychiatrists decreased from 0.56 to 0.51 visits per year per user (t = 2.98, p = 0.003), while mean payments to OMH providers rose from $62 to $69 (t = 2.46, p = 0.01). Similar to the demonstration areas, use of other physicians for mental health care in the non-demonstration areas also increased, with mean visits rising from 0.96 to 1.02 visits per year per user (t = 2.78, p = 0.005) and mean payments rising from $92 to $108 per year per user (t = 3.84, p = 0.0001). A comparison of the mean visits and payments with providers across areas reinforces the trend seen in the provider-group comparisons above—that those receiving care in the non-demonstration areas were less likely to see any mental health provider and more likely to see a non-mental health physician than their counterparts in the demonstration areas.

Effect of the Demonstration

The pre-demonstration versus post-demonstration versus control study design is intended to isolate the effect of the demonstration on mental health care utilization and expenditures by allowing one to compare pre- versus post-demonstration differences across the demonstration and non-demonstration catchment areas. However, while the non-demonstration areas were chosen to be as comparable to the demonstration areas as possible, they differ significantly from the demonstration areas in the pre-demonstration period in several important ways. For example, compared with those in the non-demonstration areas, eligible beneficiaries in the demonstration areas were more likely to be male, younger than 65, and dependents of retirees. As shown in Table 5.3, eligible beneficiaries in the demonstration areas (during the pre-demonstration period) were also more likely to have seen a counselor, psychiatrist, or OMH provider and less likely to have seen only a primary care physician for their mental health care.

To control for these population differences, we used propensity score weighting to adjust the non-demonstration group population for differences in age, sex, member category, and interactions between these characteristics. We used these propensity score weights to control for variation in the only personal information we had available between the populations and then compared weighted means across the two groups to test for statistically significant differences between the demonstration and non-demonstration areas. We first compared utilization across the two eligible populations, including the rate of any mental health care use or counselor use. We then compared rates of use among those seeing an LMHC. To determine if the demonstration had a significant impact on the variables of interest, we used a difference-in-difference approach to determine whether the differences between pre- and post-demonstration in the demonstration areas are significantly different from the differences between pre- and post-demonstration in the non-demonstration areas.

Table 5.4 presents the difference-in-difference analysis comparing means of the major analytic outcomes of interest (e.g., mean number of mental health care visits, mean expenditures for mental health care) from this weighted sample with means from the demonstration areas' eligible beneficiary population. As this table shows, differences in the major utilization outcomes (including total dollars spent on mental health care, number of visits, days of inpatient hospitalization, total dollars spent on outpatient care, and total dollars

Table 5.4
Difference-in-Differences Analyses, Eligible Beneficiaries in Demonstration Areas Versus Weighted Eligible Beneficiaries in Non-Demonstration Areas

Outcome Measure	Mean per Eligible Beneficiary		Weighted Control		Difference-in-Difference	SE[a]	95% Confidence Interval	
	Pre-Demonstration	Post-Demonstration	Pre-Demonstration	Post-Demonstration			Lower	Upper
Total mental health care dollars	$110.28	$136.16	$83.19	$108.30	$0.77	$12.45	$(23.64)	$25.17
Total outpatient mental health care dollars	$29.44	$35.04	$18.78	$22.89	$1.49	$1.65	$(1.75)	$4.73
Total mental health outpatient visits	0.34	0.37	0.27	0.30	(0.00)	0.01	(0.02)	0.02
Total LMHC visits	0.04	0.05	0.02	0.03	0.00	0.00	(0.01)	0.01
Total inpatient mental health care dollars	$80.84	$101.12	$64.40	$85.40	$(0.72)	$12.22	$(24.68)	$23.23
Total inpatient mental health care days	0.16	0.19	0.15	0.19	(0.00)	0.02	(0.03)	0.03
Percent with any mental health care use	9.26%	10.11%	9.26%	10.03%	0.09%	0.16%	(0.22%)	0.41%
Percent with any inpatient stays	1.27%	1.49%	1.36%	1.43%	0.15%*	0.06%	0.03%	0.26%
Percent with any LMHC visits	0.45%	0.55%	0.29%	0.34%	0.05%	0.03%	(0.01%)	0.12%
Percent with any psychiatrist visits	1.32%	1.45%	1.50%	1.52%	0.10%	0.06%	(0.02%)	0.22%
Percent with any OMH visits	2.02%	1.96%	1.48%	1.56%	(0.13%)*	0.07%	(0.26%)	0.00%
Percent with any mental health care visits to non–mental health care providers (other physicians)	3.94%	4.38%	4.00%	4.66%	(0.22%)*	0.10%	(0.42%)	(0.02%)
Total psychotropic drugs	0.12	0.13	0.12	0.12	0.005	0.003	(0.001)	0.010
Percent taking any psychotropic drug	6.34%	6.79%	6.50%	6.86%	0.09%	0.13%	(0.17%)	0.35%

NOTES: *Significant at the p < .05 level; ( ) denotes a negative number.
[a]Standard errors (SE) were calculated using pooled variance.

spent on inpatient care) were not significant at the 95-percent confidence level between the demonstration and non-demonstration areas. Only a few changes in outcome measures were significant at the 95-percent confidence level.

Beneficiaries in the demonstration areas were significantly less likely in the post-demonstration period to see a mental health services provider other than an LMHC or psychiatrist, and were also less likely to see a non-psychiatrist physician for mental health care. The percentage of people seeing an LMHC in the demonstration areas also increased, and although the change was not quite significant at the 95-percent confidence level, the combination of these three outcomes suggests that the demonstration may have resulted in a shift in users accessing LMHCs rather than other providers of mental health care (i.e., a substitution effect). Finally, although mean days in a hospital and mean costs for inpatient mental health care did not change significantly, users in the demonstration areas were slightly more likely to be hospitalized in the post-demonstration period than users in the non-demonstration areas. The slightly increased likelihood of inpatient mental health care in the purchased care setting among users in the demonstration areas was not offset by an increased use of inpatient mental health care in the direct care system among users in the non-demonstration areas. When examining direct care system use to investigate a potential offset, we found a decrease in inpatient mental health services use in the direct care system for both the demonstration and non-demonstration areas.

Because the demonstration changed only the rules for accessing an LMHC, we expect that any demonstration effect would be concentrated in the population most likely to see an LMHC. We therefore created a second set of weights for mental health services users in the non-demonstration areas to reflect each individual's similarity to those who saw an LMHC in the demonstration areas. Ideally, in creating these weights, we would have adjusted for the clinical characteristics of mental health care users, including diagnoses and possibly the use of psychotropic medications. However, we expect that the recording of diagnoses on claim records, as well as the prevalence of the number and types of medications prescribed, might vary based on the type of provider an individual saw (based on the traditional treatment orientations of the various provider groups, even given the same reasons for visits or underlying needs for mental health care). For example, we expect that mental health diagnoses are less likely to be recorded on a primary care physician's records than they would be on a psychiatrist's records. We therefore matched only on main demographic characteristics—age, sex, and member category. The small sample size also prevented us from using interaction terms to create this set of weights.

Table 5.5 compares the weighted non-demonstration population with the group of those who saw an LMHC in the demonstration areas. Comparing Table 5.5 to Table 5.3, we note that while the weighted non-demonstration population has almost twice the rate of LMHC use as the unweighted control group population, it still has a low rate of LMHC use (weighted population: 5.64 percent at pre-demonstration and 6.19 percent at post-demonstration; unweighted: 3.0 percent at pre-demonstration and 3.1 percent at post-demonstration). Table 5.5 shows that the only outcome change that is significantly greater at the 95-percent confidence level in the demonstration areas is the probability of seeing a psychiatrist—that is, those in the demonstration areas seeing an LMHC were less likely to also be seeing a psychiatrist in the post-demonstration period. This reduction in the use of psychiatrists' services could potentially be a result of the removal of the physician oversight

Table 5.5
Difference-in-Differences: Demonstration Mental Health Services Users Versus Weighted Non-Demonstration Mental Health Services Users

Outcome Measure	Mean per Eligible Beneficiary				Difference-in-Difference	SE[a]	95% Confidence Interval	
			Weighted Control				Lower	Upper
	Pre-Demonstration	Post-Demonstration	Pre-Demonstration	Post-Demonstration				
Total mental health care dollars	$1,504.33	$1,349.49	$979.43	$1,250.96	$(426.37)	$323.66	$(1,060.74)	$208.00
Total outpatient mental health care dollars	$802.16	$749.46	$317.65	$357.46	$(92.51)	$66.83	$(223.50)	$38.47
Total mental health outpatient visits	12.96	12.25	4.32	4.34	(0.73)	0.71	(2.13)	0.67
Total LMHC visits	9.24	8.54	0.44	0.48	(0.74)	0.57	(1.85)	0.38
Total inpatient mental health care dollars	$702.16	$600.03	$661.78	$893.50	$(333.86)	$311.17	$(943.75)	$276.03
Total inpatient mental health care days	0.71	0.70	0.74	0.84	(0.11)	0.24	(0.58)	0.36
Percent with any inpatient stays	9.45%	7.47%	11.36%	10.64%	(1.26%)	1.59%	(4.37%)	1.85%
Percent with any LMHC visits	100.00%	100.00%	5.64%	6.19%	(0.54%)	0.34%	(1.21%)	0.12%
Percent with any psychiatrist visits	41.63%	32.13%	24.23%	22.88%	(8.14%)*	2.70%	(13.44%)	(2.85%)
Percent with any OMH visits	21.72%	22.80%	28.79%	27.42%	2.45%	2.36%	(2.18%)	7.08%
Percent with any mental health visits to non–mental health care providers (other physician)	24.54%	25.20%	39.99%	43.06%	(2.42%)	2.46%	(7.23%)	2.40%
Total psychotropic drugs	2.01	1.53	1.41	1.28	(0.36)*	0.11	(0.56)	(0.15)
Percent taking any psychotropic drug	73.30%	62.53%	70.09%	66.33%	(7.01%)*	2.60%	(12.11%)	(1.91%)

NOTES: *Significant at the p < .05 level; ( ) denotes a negative number.
[a]SEs were calculated using pooled variance.

requirement *if* LMHCs had previously been co-treating beneficiaries with psychiatrists as a means of fulfilling the supervision requirement and then stopped doing so when the supervision requirement was removed. While the changes are not significant at the 95-percent confidence level, the drop in the likelihood of seeing a non–mental health physician and the drop in the mean number of mental health visits per user also support the hypothesis that those seeing an LMHC were less likely to also get care from a physician as a result of the demonstration. Furthermore, the decreases in the likelihood of using a psychotropic medication and the mean number of prescriptions for psychotropic drugs per person (see Table E.7) are significant in the weighted difference-in-difference comparison, indicating that the demonstration may have decreased the prevalence of psychotropic drug use among people seeing a counselor.

We were concerned about the low levels of LMHC use in the comparison sample, as shown in Table 5.5. We therefore repeated the propensity score weighting, this time including only control group users who saw an LMHC, as a sensitivity analysis. We once again matched on age, sex, and member category. This difference-in-difference comparison of LMHC users is presented in Table 5.6. As expected, the mean number of visits per user is much higher than in the previous analysis. As in the previous analysis, mental health care users in the demonstration areas were significantly less likely to see a psychiatrist and had fewer psychotropic drug claims in the post-demonstration period. The likelihood of having any psychotropic drug claim also fell, although the effect was not significant at the 95-percent confidence level.

In summary, the demonstration appeared to impact utilization in the following ways: Among the entire eligible beneficiary population in the demonstration areas, there was an increase in the likelihood of having an inpatient hospitalization, a decrease in the likelihood of seeing an OMH provider, and a decrease in the likelihood of seeing a non–mental health services provider (other physician) for mental health care. Changes in inpatient and outpatient costs were small and not statistically significant. Further refinement of the difference-in-difference analyses to control for differences in the characteristics of those who see LMHCs revealed a significant decrease in the likelihood of seeing a psychiatrist and a decrease in the likelihood of receiving a psychotropic drug.

Unfortunately, based on administrative data alone, it is not possible to determine whether these changes had a clinically significant impact on beneficiaries. While the increase in the likelihood of inpatient hospitalization over the entire eligible beneficiary population is of some concern as a potential measure of quality of care, the fact that the rate of hospitalization did not increase in the LMHC group suggests that the increase may have had some cause other than the demonstration. Also, while the demonstration did appear to impact the type and source of care beneficiaries received, we cannot ascertain whether being less likely to see a physician and receive a psychotropic medication had any impact on the clinical outcomes for these individuals. While we did seek to examine whether a clinically relevant change could be observed in adverse events, such as suicide attempts, the type of data available for this study are not ideal for such analyses. For example, we found zero occurrences of visits to emergency departments in the purchased care sector for injuries sustained as a result of a suicide attempt. This result does not necessarily mean there were no such attempts, rather that they are not necessarily coded in the claims data. We also looked at the direct care system data to evaluate the occurrence of suicide attempts. Codes for such injuries in this

Table 5.6
Difference-in-Differences: Demonstration Mental Health Services Users Versus Weighted Non-Demonstration Areas

| Outcome Measure | Mean per Eligible Beneficiary | | | | Difference-in-Difference | SE[a] | 95% Confidence Interval | |
| | Pre-Demonstration | Post-Demonstration | Weighted Control | | | | Lower | Upper |
			Pre-Demonstration	Post-Demonstration				
Total mental health care dollars	$1,504.33	$1,349.49	$1,085.37	$1,465.83	$(535.30)	$350.34	$(1,221.95)	$151.36
Total outpatient mental health care dollars	$802.16	$749.46	$668.86	$700.88	$(84.73)	$83.48	$(248.34)	$78.89
Total mental health outpatient visits	12.96	12.25	10.85	10.74	(0.60)	0.94	(2.43)	1.24
Total LMHC visits	9.24	8.54	7.55	7.44	(−0.59)	0.75	(2.06)	0.89
Total inpatient mental health care dollars	$702.16	$600.03	$416.52	$764.95	$(450.57)	$331.15	$(1,099.63)	$198.49
Total inpatient mental health care days	0.71	0.70	0.64	0.96	(0.32)	0.31	(0.93)	0.28
Percent with any inpatient stays	9.45%	7.47%	10.84%	11.42%	(2.56%)	2.36%	(7.18%)	2.07%
Percent with any LMHC visits	100.00%	100.00%	100.00%	100.00%	0.00%	0.00%	0.00%	0.00%
Percent with any psychiatrist visits	41.63%	32.13%	40.94%	40.71%	(9.27%)*	3.86%	(16.83%)	(1.70%)
Percent with any OMH visits	21.72%	22.80%	15.26%	15.37%	0.97%	3.07%	(5.04%)	6.98%
Percent with any mental health visits to non–MH providers (other physician)	24.54%	25.20%	25.97%	27.56%	(0.93%)	3.47%	(7.74%)	5.88%
Total psychotropic drugs	2.01	1.53	1.73	1.64	(0.40)*	0.15	(0.69)	(0.11)
Percent taking any psychotropic drug	73.30%	62.53%	70.55%	66.79%	(7.01%)	3.67%	(14.19%)	0.18%

NOTES: *Significant at the p < .05 level; ( ) denotes a negative number.
[a]SEs were calculated using pooled variance.

data were in fact very rare, and the very low percentage (less than 0.01 percent) in the demonstration group and the non-demonstration groups were not significantly different.

Impact on Administrative Costs Associated with Referral and Supervision

The FY01 NDAA legislation requested a description of the administrative costs associated with referral and supervision requirements under TRICARE. At the outset, however, it is worth noting that a full description of the administrative costs of LMHC referral and supervision requirements necessitates identifying the bearers of such costs. Costs may accrue for several reasons. The completion of paperwork related to those requirements would undoubtedly create some administrative costs for LMHCs, but the requirements could also create administrative costs for other clinical providers (in their roles as supervisors), for TRICARE managed care contractors, or for TRICARE itself (e.g., in auditing compliance by contractors with the requirements). It is reasonable to expect that there is a cost associated with the time required for LMHCs and those supervising them to fulfill these requirements. Note, however, that referral and supervision are not billable services and, as such, neither LMHCs nor the physicians who might refer beneficiaries to and supervise those LMHCs (and the referring and supervising physician may not necessarily be the same individual) can bill TRICARE for the time associated with meeting these requirements. Consequently, the administrative costs associated with meeting and documenting these requirements are not easily quantified.

Note also that some of the potential costs of referral and supervision requirements for LMHCs may be subtle. In particular, to the extent that the requirements create disincentives for beneficiaries to seek care from LMHCs, the result might be to reduce the demand for LMHCs' services. In a sense, lost patronage for LMHCs could be viewed as an administrative cost associated with the referral and supervision requirements. Substitution of demand for mental health services toward higher-cost providers might also be construed as a related administrative cost. We do not address these forms of administrative costs here.

To investigate the administrative costs to TRICARE's MCSCs associated with the referral and supervision requirements for LMHCs, we interviewed MCSC officials in both the demonstration areas and non-demonstration areas. Moreover, for the MCSCs that actually participated in the demonstration, we engaged in two sets of interviews, at the beginning and at the end of the demonstration period. In each of these interviews, we asked respondents a series of questions concerning the administrative requirements for LMHCs under TRICARE, the administrative costs to the MCSCs in enforcing those requirements, and any advantages or disadvantages accruing to independent practice by LMHCs (i.e., from the MCSC's perspective).

In general, the representatives from all three of the MCSCs that participated in our study (one MCSC for the demonstration area and two that covered the non-demonstration areas) agreed that the pre-demonstration administrative requirements for LMHCs under TRICARE included physician referral and supervision. All agreed that the referral requirement is burdensome primarily to the LMHCs themselves and to beneficiaries, by imposing a barrier to patients seeing LMHCs for care, and is an incentive for patients to seek therapy from other types of providers.

The MCSC respondents actually differed in their description of what the baseline supervision requirement entails, likely the result of differences in how each of the MCSCs implements and enforces the supervision policy. For example, one of the respondents from a non-demonstration MCSC said that LMHCs in that area were required to simply provide the name of a supervising physician on a periodic Treatment Authorization Request form,[6] that no signature was ever required from the supervisor, and that no major administrative costs to the MCSC were associated with supervision (hence, no likely savings from removal of the requirement). By contrast, a respondent from the other non-demonstration MCSC said that LMHCs must show a "documented ongoing relationship" with a supervising physician, that clinical proof of supervision is required for every eight therapy visits, and that these requirements are extremely burdensome for LMHCs to meet. Moreover, this respondent also said that these requirements were burdensome for the MCSC and that associated costs from paperwork and time resulted in LMHCs being about 25 percent more expensive for them to manage than other types of providers.

Respondents from the demonstration MCSC offered still another perspective on the supervision requirement. They reported that LMHCs were required to have their treatment notes signed by their supervisors, but that actual enforcement of supervision occurred mostly through the filing of claims forms (on which a supervisor's name had to be included). With regard to associated administrative costs, the respondents suggested that removal of the supervision and referral requirements would eliminate some paperwork for the MCSC and could result in a slight improvement in administrative efficiency. However, following the demonstration, the same respondents indicated that there was little or no change in their own administrative costs as a result of removing the supervision and referral requirements. The demonstration MCSC respondents also said that, to the best of their knowledge, there was no indication of any change in the nature or quality of care delivered by LMHCs during the demonstration (e.g., there had been no adverse events or complaints made against participating LMHCs during the course of the demonstration period).

The consistent theme that emerged from our interviews with MCSC officials was that the perceived advantage of the demonstration (i.e., the perceived advantage of independent practice for LMHCs) did not manifest itself in reduced administrative costs to MCSCs but rather in increased access to therapy services for TRICARE beneficiaries. Several of the interview respondents acknowledged that the referral and supervision requirements for LMHCs under TRICARE may make it harder for beneficiaries to see these providers, while creating an incentive for beneficiaries to seek out other types of mental health service providers (social workers, psychologists, psychiatric nurse specialists). Our MCSC respondents were divided about whether independent practice for LMHCs might result in quality-control problems, in part due to the existence of heterogeneous licensing standards for mental health counselors across different states within the United States. Even those respondents who expressed this concern, however, suggested that improved credentialing standards for counselors would be a more effective way to safeguard beneficiaries and to promote the quality of care overall for those who seek care from mental health counselors.

---

[6] A therapist is reportedly required to submit a Treatment Authorization Request for every eight therapy visits to obtain continuing reimbursement for that patient under TRICARE.

Implications of Findings, Caveats, and Conclusions

Implications of Findings

The study findings presented in this report have several important implications for TRICARE. The data presented in this report provide a unique picture of mental health service use within the TRICARE beneficiary population. Although the study was limited to only six catchment areas, the results provide a glimpse of the characteristics of TRICARE beneficiaries who use mental health services and describe the utilization patterns and costs associated with the delivery of mental health services to this special population. The results also provide interesting insights into beneficiaries' need for, perceptions of, and satisfaction with mental health service use. More specifically, our survey data contribute significantly to the mental health services and military health care fields, given that no other survey has looked at a TRICARE beneficiary group that consists exclusively of documented consumers of mental health services. Other surveys have examined the perceived impact of military life on active duty personnel (Bray et al., 2003); however, this is the only independent study that we know of to examine mental health symptoms and other factors related to mental health service use among family members of active duty personnel and among retirees and their family members. Based on our survey, we found little impact by demonstration area on utilization of mental health care services. However, consistent with our hypotheses, we did find that the perceived social stigma associated with seeking mental health care for military health beneficiaries was connected with lower mental health care utilization and higher rates of medication use over and above the effect of the demonstration.

Recent publicity, including a 2004 article in the *New England Journal of Medicine* (Hoge et al., 2004) and articles in the lay press, has focused attention on mental health problems of military personnel and the potential need for more mental health services within the military population. Use of mental health services may be high among military family members and retirees, particularly during the present wartime situation. Because a significant proportion of TRICARE mental health users are spouses of active duty military members or are retirees with adult children serving in active duty status, greater attention to family needs during deployments may aid these beneficiaries in coping with mental health–related symptoms. These factors provide a compelling reason to learn about the mechanisms that impede the use of mental health services. Although this study was structured as an evaluation of independent practice for LMHCs under TRICARE, our findings offer insights into broader issues concerning access and service use during wartime and can help guide policymakers toward strategies to improve access to TRICARE mental health services.

Study Limitations and Caveats

Several limitations and caveats should be noted in interpreting our findings. These include the initial selection of the demonstration areas, constraints associated with the type of data that are required and available for our analysis, restrictions against some eligible beneficiaries for demonstration participation, and the focus on only the purchased care system within TRICARE.

In choosing the demonstration areas, TMA first selected the TRICARE health care region with the highest absolute number of visits to mental health counselors in FY00—the Central Region (at the time identified as "region 7/8" and managed by TriWest). Then, TMA selected the catchment areas that had the greatest number of mental health counselors relative to the other catchment areas in the Central Region. TMA made this selection to guarantee that enough beneficiaries would be included under the demonstration to provide ample statistical power for analyses of claims data as well as for a potential survey. However, to better test whether this demonstration expanded or improved access to LMHCs, a region in which mental health counselors were not already heavily utilized perhaps would have been more informative. In turn, the demonstration area selection methodology restricted the selection of suitable comparison areas to those areas in which LMHCs were already being utilized at similar rates. This ruled out consideration of the Upper Northwest, where visits to LMHCs accounted for less than 1 percent of all mental health visits for FY00.

We were limited by the type of data available to us to perform the study. Because we had to rely on the use of pre-existing claims data, our analyses were based primarily on currently available variables. In most cases, these variables are recorded for purposes other than assessment of mental health service utilization and treatment process outcomes. As such, the validity of our measures depended on the validity of the information recorded in the claims. The analyses were also limited to mental health users in the purchased care sector (contracted care). Beneficiaries who use only direct care services (i.e., care received in a military-owned treatment facility) for mental health treatment were not included in our analyses.

Limitations on the survey of beneficiaries should also be noted. First, a cross-sectional survey does not allow for fully adjusting for pre-existing differences between groups prior to the demonstration. Although the claims data were available to adjust at the aggregate level, we were unable to match individual-level data because of concerns regarding HIPAA. While this could have affected our findings, we minimized potential bias by weighting the sample for non-response. Age was the only significant predictor of non-response in this sample, and weighted analyses account for this bias. Second, the survey responses relied on self-reported data. As with any self-reported data, responses may be subject to recall bias and selection of socially desirable responses. However, we employed mostly established measures that have been widely used and validated in previous studies, which minimizes any bias. Moreover, the use of self-reports for understanding the patient's or beneficiary's perspective about health circumstances is believed to be the most appropriate method, because it is the subjective report that matters the most.

Finally, the generalizability of our findings is limited based on the restriction on the involvement of mental health providers who practice in the purchased care system, and the findings are based only on the care they render to MHS beneficiaries over the age of 18 years. Because LMHCs treat primarily non–active duty beneficiaries who receive care in the purchased care system, those beneficiaries who receive all of their health care in the direct care system (e.g., much of the active duty population) were likewise not exposed to the demonstration. As such, we cannot assess whether or not independent practice authority for LMHCs (i.e., the demonstration) provided expanded access to mental health services, or expanded access to LMHCs more specifically, for beneficiaries under the age of 18 or the general active duty population, two groups for whom there may be concerns about adequate mental health services support within the MHS (Hoge et. al, 2004; Bray et al., 2003).

Conclusions

In summary, we found that the evaluation of the DoD Mental Health Counselor Demonstration for expanded access to LMHCs under TRICARE had minimal impact on the variety of outcomes we studied. Access to mental health care, as measured by the percentage of eligible beneficiaries who used mental health services, increased in both the demonstration and non-demonstration areas. Most of the increase is probably due to the fact that the demonstration coincided with the beginning of the Iraq War, rather than any increased perception among potential beneficiaries of expanded access to mental health care. In addition, there were no key effects on expenditures, reimbursement, administrative costs, or patient confidentiality. While we did see increases in utilization and costs for mental health care over the demonstration period, these increases could not be attributed to allowing independent practice authority. In fact, according to the annual *Evaluation of the TRICARE Program* report (Institute for Defense Analyses et al., 2004), both utilization and costs of health care services increased for the overall TRICARE population during the same time period.

Using TRICARE administrative claims data, we found that the demonstration did likely impact the type of providers from whom beneficiaries sought mental health care and the likelihood of users receiving a psychotropic medication. Specifically, among the eligible population, there was a decrease in the likelihood of seeing an OMH provider, a decrease in the likelihood of seeing a non–mental health physician (other physician) for mental health care, and an increase in the likelihood of having a mental health inpatient hospitalization (that was not offset by utilization of inpatient mental health services in the direct care system). Changes in inpatient and outpatient costs were small and not statistically significant. Further refinement of the difference-in-difference analyses to control for differences in the characteristics of those who see LMHCs revealed a significant decrease in the likelihood of users seeing a psychiatrist and a decrease in the likelihood of their receiving a psychotropic drug. However, based on administrative data alone, it is not possible to determine whether these changes had a clinically significant impact on beneficiaries.

Where we did observe some potential positive effects was in ratings of satisfaction. According to self-reported survey data from beneficiaries, those in the demonstration areas reported higher ratings of satisfaction with their mental health services than those in the non-demonstration areas; however, due to the cross-sectional nature of the data, it is not possible to determine if differences in ratings of satisfaction also existed prior to the demonstration. The effect on administrative costs associated with the requirements for LMHCs was also unclear. From our interviews with LMHCs and other mental health services providers, it was apparent that supervision and referral were not that onerous to begin with, and that any administrative costs associated with the requirements were in fact minimal at the outset.

Lastly, the effectiveness of mental health care provided by LMHCs versus other mental health services providers could not be estimated due to the lack of clinically relevant data on mental health care users. Such analyses are possible only when patients can be tracked over time in order to measure the impact and adequacy of the treatments received. Because the current TMA privacy requirements did not allow us to collect data in this manner, it was not possible to estimate the effects of the demonstration on the quality of care provided to beneficiaries.

Table 6.1 summarizes the key findings and implications for each of the legislative objectives for this evaluation that were mandated by Congress. Taken as a whole, our findings suggest that the impact of expanding access to LMHCs for providers and beneficiaries on beneficiaries, providers, and the TRICARE program was minimal. Nevertheless, the findings are important in the sense that they indicate that merely lifting administrative requirements for the provision of mental health care by itself is unlikely to result in expanded access and utilization, especially when beneficiaries already have access to other types of mental health providers who do not have the same administrative requirements as the LMHCs but can provide many similar services. These findings suggest that reducing the social stigma attached to mental health care and expanding access to mental health care must go beyond merely lifting the administrative requirements on LMHCs.

Table 6.1
Summary of Evaluation Findings and Implications for Each Legislative Objective

Legislation Objective	Key Findings	Implications
1. Describe the extent to which expenditures for LMHCs changed as a result of allowing independent practice	Controlling for beneficiary characteristics, there was no significant change in expenditures for inpatient and outpatient care among the eligible population or among those seeing LMHCs.[a]	Allowing for increased access to LMHCs had no measurable impact on expenditures for mental health services for those who received care from LMHCs.
2. Provide data on utilization and reimbursement for non-physician MH professionals	Among those MH users in the other mental health (OMH) provider group, the mean number of visits increased in both the demonstration and non-demonstration areas. [a] For those in the OMH group, total expenditures for MH care increased in both the demonstration and non-demonstration areas. Comparing the changes pre- and post-demonstration and demonstration versus non-demonstration, we found a decrease in the likelihood of beneficiaries seeing an OMH provider in the demonstration areas.	Opening up access to LMHCs may have created a substitution effect—that is, beneficiaries were less likely to see other non-physician mental health providers, such as psychologists, social workers, and psychiatric nurse practitioners.
3. Provide data on utilization and reimbursement for physicians who make referrals to and supervise LMHCs	Among those MH users in the group of users who saw a psychiatrist, there were no significant changes in the mean number of outpatient MH visits in the demonstration areas or the non-demonstration areas.[a] For those MH users in the non-psychiatrist physician group, there was a statistically significant increase in the mean number of outpatient visits in the non-demonstration areas but not the demonstration areas. [a] Mean expenditures for MH care among MH users in the psychiatrist and other physician groups increased from pre-demonstration to post-demonstration in both the demonstration and non-demonstration areas, but only the increase in the non-psychiatrist "other" physician group in the non-demonstration physician area was statistically significant. Comparing the changes pre- versus post-demonstration and demonstration versus non-demonstration, we found a significant decrease in the likelihood of beneficiaries seeing a physician (psychiatrist or other physician) for MH care in the demonstration areas.	Removing the referral and supervision requirements significantly decreased the likelihood that beneficiaries would get MH care from a physician (psychiatrist or other physician) and, as such, decreased the likelihood that they would also get a psychotropic medication to treat their mental illness.
4. Describe administrative costs incurred as a result of documenting referral and supervision	According to the LMHCs we interviewed, eliminating the physician referral requirement saves time previously spent in telephone consultations to obtain supervision, confirm referrals, and authorize therapy.	The demonstration probably resulted in modest cost savings to LMHCs in terms of time and administrative burden. Any savings to MCSCs depended on their baseline enforcement procedures regarding supervision and referral (which was minimal in some cases).

Table 6.1—Continued

Legislation Objective	Key Findings	Implications
5. Compare effect for items outlined in objectives one through four, over one year (pre-post) in the demonstration areas as compared with non-demonstration areas [b]	All findings listed above are based on analyses that compared data gathered from one year prior to the demonstration with data gathered one year following the demonstration in both the demonstration and non-demonstration areas.	Not applicable
6. Describe the ways in which independent practice affects the confidentiality of MH and substance abuse services for TRICARE beneficiaries	There was no evidence that eliminating the referral and supervision requirements would change the standards for confidentiality.	Independent reimbursement of LMHCs would have no impact on confidentiality.
7. Describe the effect of changing reimbursement policies on the health and treatment of TRICARE beneficiaries	There was no effect on perceived access to MH services. There was no effect on self-reported adherence to MH treatment. There was no effect on self-reported MH status. There was a potential positive effect on HEDIS ratings of mental health services; however, positive ratings may have also been evident prior to the demonstration.	Increased access to LMHCs had no adverse effect on TRICARE beneficiaries and may be associated with greater satisfaction with MH services.
8. Describe the effect of DoD policies on the willingness of LMHCs to participate as health care providers in TRICARE	Lack of independent practice authority for LMHCs was viewed as a disincentive or barrier to participation prior to the demonstration. Demonstration participation increased initially and leveled off around the middle of the demonstration period. Enrollment of LMHCs as TRICARE network providers increased during the demonstration period, but this is not likely the result of the changing practice authority because this was a temporary demonstration.	The findings suggest that the demonstration may have been a motivator to network participation (although we have no data on network enrollment for the non-demonstration catchment areas during the same time period to use for comparison).
9. Identify any policy requests or recommendations regarding LMHCs made by TRICARE plans or managed care organizations	Removal of the referral and supervision requirements for LMHCs remains a top legislative priority for the ACA and AMHCA. According to MCSC representatives, quality concerns could be addressed by development of appropriate and standardized credentialing mechanisms.	Adoption of formal credentialing standards could help to facilitate independent practice for counselors in states with rigorous licensing, while helping to promote the implementation of similar standards elsewhere.

[a] We created hierarchical groups of users by provider type to compare differences in the changes in users' utilization patterns (see Chapter Five).

[b] Item 5 was included in the legislation as a means of describing the methods to be used for responding to objectives 1 through 4. Although it is not included as an objective in the bulleted list at the top of this summary, we include it here for consistency with the legislation.

Demonstration Materials

PARTICIPATION AGREEMENT

TRICARE Expanded Access to Mental Health Counselors Demonstration Project

This Participation Agreement ("Agreement") is between the United States of America through the Department of Defense, TRICARE Management Activity ("TMA"), a field activity of the Office of the Secretary of Defense, the administering activity for the TMA and _____ ("Provider").

The purpose of this participation agreement is to:

a. Establish the Provider's participation in the TRICARE Expanded Access to Mental Health Counselors Demonstration Project ("Demonstration").

b. Establish the terms and conditions of the Provider's participation in the Demonstration.

SECTION 1

General Agreement

1.1 TMA agrees to waive the TRICARE requirements for the Provider to have physician referral and supervision during the demonstration period. TRICARE contractors will be instructed to pay claims of participating Providers accordingly.

1.2 The demonstration period will begin on January 1, 2003 or the execution date of this Agreement, whichever is later. The demonstration period will end December 31, 2004.

1.3 TMA, or its designee, will analyze aggregated data collected from claims and other available sources to evaluate the impact of independent reimbursement of mental health services provided by selected mental health counselors.

SECTION 2

Provider Requirements

1. Provider agrees to collect the TRICARE Mental Health Counselors Demonstration Project Informed Consent Form from all TRICARE patients during the demonstration period. The form informs the TRICARE member that the Provider is participating in the TRICARE Mental Health Counselor Demonstration, which allows the Provider to provide services to the TRICARE member without physician referral or supervision.

2. Provider agrees to keep Merit Behavioral Care's TRICARE Central Region Office ("MBC TRICARE") notified of any address, telephone, or tax identification number changes. Changes can be sent to the MBC TRICARE fax line at 1-602-564-2336.

3. Providers should send Demonstration-related documents and correspondence to the fax line cited above or to MBC TRICARE, P.O. Box 42150, Phoenix, AZ 85080-2150. Providers may also call the MBC TRICARE Provider Relations line at 1-888-910-9378 for assistance.

4. Provider agrees that there will be no additional compensation for participating in the Demonstration.

SECTION 3

Termination and Amendment

3.1 TMA may terminate this Agreement with 30 days written notice if the Demonstration is cancelled.

3.2 This Agreement will terminate immediately if a provider relocates outside of the Offutt AFB catchment area, Ft. Carson catchment area, or USAF Academy catchment area.

3.3 The Executive Director, TMA, or designee, may amend the terms of this Agreement by giving 30 days notice in writing of the proposed amendment(s).

3.4 Either party may terminate this Agreement without cause upon 30 days written notice of termination to the other party.

SECTION 4

Effective Date

This Agreement is effective on the date signed by the Executive Director, TMA, or designee.

TMA   **PROVIDER**

Signature: _____   Signature: _____

Printed Name: _____   Printed Name: _____

Executed on _____, 20__

Informed Consent Form

Research Study
TRICARE Mental Health Counselors Demonstration Project

INTRODUCTION
You are being asked to take part in a research study. Before you decide to be a part of this research study, you should read the information below and need to understand it so that you can make an informed decision. This is known as informed consent.

PURPOSE AND PROCEDURES
The TRICARE Management Activity, through the Department of Medical and Clinical Psychology of the Uniformed Services University of the Health Sciences, is carrying out a congressionally mandated demonstration project to study the effects of waiving the requirements for mental health counselors to receive their referrals from a physician and to receive ongoing supervision from a physician. Under current TRICARE regulations, licensed or certified mental health counselors are required to document that a physician has referred TRICARE beneficiaries whom they treat. They are also required to receive ongoing supervision of their services by a physician. For the purposes of this demonstration project, counselors have independent practice authority. This means that your provider will not be receiving ongoing case supervision by a physician. At the end of the project, TRICARE will make comparisons between beneficiaries who received services from counselors with other types of providers. You might be asked to voluntarily participate in an optional survey concerning the quality of your care. However, your responses would be kept completely confidential, and no one, <u>not even your counselor,</u> would have access to any feedback you provide.

POSSIBLE BENEFITS
By participating in this study, you may be expanding the range of mental health providers available to you. Possibly, counselors who would not otherwise consider becoming TRICARE providers would now do so.

POSSIBLE RISKS
Mental health counselors are ordinarily required to be medically supervised under TRICARE. Your provider, as a participant in this demonstration project, is granted independent practice authority and will not be medically supervised. However, he/she will promptly refer any medical concerns or referrals for medication evaluation to a physician should circumstances require it.

ALTERNATIVES
If you do not wish to receive services from a Mental Health Counselor Demonstration provider, you may call 1-888-910-9378 for a referral to another mental health provider.

COSTS
There are no additional costs associated with participating in this demonstration project.

RIGHT TO WITHDRAW FROM THE STUDY
Your participation in this research study is completely voluntary. You may decide to stop taking part in this study at any time by terminating your professional relationship with this provider. You may then seek an alternative provider by calling the telephone number cited above.

PRIVACY

As always, your medical records are kept by your provider and are never shared with anyone else. If you are asked to complete a survey, any information you provide will have any identifying information removed, and all responses will be combined with all other program participants, so that your privacy will be guaranteed. Again, **your individual identifying information will never be made available to anyone.**

QUESTIONS

If you have any questions about this project, you should contact CAPT Mark Paris at (703) 681-0064. If you have any questions about your rights as a research subject, you should call the Director of Research Programs in the Office of Research at the Uniformed Services University of the Health Sciences at (301) 295-3303. This person is your representative and has no connection to anyone conducting the study.

SIGNATURES

By signing this consent form, you are agreeing that the study has been explained to you and that you understand the study. You are signing that you agree to take part in this study. You will be given a copy of the consent form.

Signature: _____   Witness Signature: _____

Date: _____   Date: _____

COUNSELOR STATEMENT

I certify that this project has been explained to the above individual, by me or my staff, and that the individual understands the nature and purpose, the possible risks and benefits associated with taking part. Any questions that have been raised have been answered.

_____   _____

Mental Health Counselor/staff member   Date

Evaluation Tasks And Methods

The study was organized into four tasks. The design of three of those tasks was based on the source of data. This appendix details the objective of each task and the methodology and analyses employed for each.

Task 1
Review and provide feedback on demonstration plans to determine suitability for evaluation purposes

Objectives: To provide feedback to DoD on the suitability of the implementation plans for evaluation purposes and to ensure the proper design and selection of methods for evaluating the impact of the LMHC demonstration.

Task Design and Procedures:

1. Provide comments on the demonstration plans, including the informed consent forms, procedures for participants and beneficiaries, and the Institutional Review Board materials.

   As requested, RAND reviewed plans, generated by the TRICARE Management Activity, for implementing the demonstration. This review included participating in conference calls with TMA, Merit/Magellan Behavioral Health, and TRIWest. In addition, and as requested, we provided information with regard to our evaluation plan/protocol to TMA so that it could submit the necessary Institutional Review Board forms for the demonstration. Throughout the task, we took great care to ensure that all feedback specifically focused on our own ability to evaluate the impact of the demonstration given the implementation protocol. As such, we did not give any formal guidance or suggestions on how the implementation protocol should be designed or launched.

2. Obtain preliminary estimates on the number of providers and beneficiaries in the demonstration areas for purposes of creating a sampling plan.

   To inform the process of creating sampling plans and budget estimates for the beneficiary survey (described in Task 3 below), RAND requested and received rough analyses of the total number of visits (and unique number of beneficiaries making up those visits) to different mental health providers (mental health counselors, psychologists, social workers, psychiatrists, psychiatric nurses, pastoral counselors) in each of the selected demonstration catchment areas. In addition, we requested and received counts of the number of counselors in each catchment area (to estimate the number of beneficiaries per counselor). The visit data were collected from the Health Care Service Record, TMA, and the initial provider data from TRIWest records. These reports were used to assess whether there would be a sufficient number of eligible beneficiaries for sampling purposes (assuming a 50-percent response rate) to ensure statistical power (see Task 3) for the main evaluation analyses.

3. Advise DoD on the selection of a non-demonstration comparison area(s) for purposes of pre- and post-demonstration analytic comparisons.

    To facilitate TMA's review and selection of comparison area(s), RAND met with the project sponsor to discuss and prioritize possible selection criteria. At this meeting, RAND proposed consideration of several possible criteria to be used in selecting comparison areas. RAND advised on selection of catchment areas for purposes of comparison that matched demonstration areas on the following characteristics:

- MTF size (based on number of providers, which potentially serves as a proxy for the availability of services on base)
- Branch of service (for MTF in catchment area; the demonstration areas included one Army and two Air Force catchment areas)
- Geographic region (TMA requested that we not consider catchment areas on the East Coast due to possible contamination in mental health service utilization surges following the September 11, 2001, attacks)
- Percentage of eligible beneficiaries in the catchment areas who used an outpatient, purchased care mental health service during the past fiscal year
- Frequency distribution of total outpatient, purchased care visits (for eligible beneficiaries 18 and over), by mental health provider
- Frequency distribution of mental health users (eligible beneficiaries 18 and over), by mental health provider
- Number and proportion of network-enrolled providers in each mental health provider group.

    TMA agreed that, among the above characteristics, the utilization patterns of visits to the various mental health provider groups were the primary criteria by which it wanted the demonstration and comparison areas to be matched. Other criteria of importance were agreed to be the number of beneficiaries who sought services from each of the provider groups and the proportional distribution of each of the mental health provider groups. On request, RAND agreed to review potentially relevant data available on the TMA Web site on outpatient mental health service utilization, receive some rough data analyses conducted by TMA, and provide feedback to TMA for its non-demonstration area selection process.

    Data Sources: Three primary sources of data and or information were reviewed:

- TRICARE Web site: For information on the branch of service and managed care support contractor, behavioral health contractor, and the Health Care Summary Reports for each of the catchment areas of interest (see, for example, http://199.211.83.250/Reports/HR/2001/default.htm)
- Health Care Service Records (summary reports provided by TMA): To generate reports on the total number of visits for beneficiaries 18 and over who sought services from mental health providers (sorted by provider type) and the corresponding number of unique beneficiaries 18 and over who used such services during FY01

- Health Care Provider Records (summary reports provided by TMA): To generate reports on the total number of network-enrolled providers in each of the provider-type categories of interest for each catchment area of interest.

To begin the extraction of information on potential catchment areas for use as comparison areas, RAND conducted a preliminary, online review of statistical reports to narrow down the number of catchment areas to be considered. We concentrated our attention on catchment areas that seemed to be similar in size and geography to the demonstration areas. More specifically, we focused on potential areas that were

- within a health care service region that had percentages of mental health counselor utilization similar to those in the demonstration areas. Visits to mental health counselors accounted for roughly 16 percent of the visits to all mental health providers in Region 7/8 during FY00. The other regions with similar proportions were Regions 2, 3, 4, 5, and 6. However, due to TMA's concerns about the surge in mental health service use in Regions 1 and 2 following the September 11, 2001, attacks, we excluded those regions from consideration.
- *a priori* believed to have comparably sized MTFs.
- *a priori* believed to be close to a mid- to large-sized metropolitan area.
- either an Army- or Air Force–managed catchment area (there is no Navy catchment area in the demonstration region, and both Army and Air Force catchment areas were considered because the organization and delivery of health care service can vary depending on the branch of service).

Using the above criteria, we selected the following catchment areas for closer evaluation: Ft. Gordon, Georgia (Army, Region 3); Ft. Bliss, Texas (Army, Region 7/8); Ft. Hood, Texas (Army, Region 6); Luke AFB, Arizona (Air Force, Region 7/8); Wright-Patterson AFB, Ohio (Army, Region 5); and MacDill AFB, Florida (Army, Region 3). We requested data from TMA on these catchment areas as well as the demonstration areas: Ft. Carson, Colorado (Army, Region 7/8); USAF Academy, Colorado Springs (Air Force, Region 7/8); and Offutt AFB, Nebraska (Air Force, Region 7/8).

After the data were extracted and tabulated, RAND collated all available estimates in an Excel spreadsheet, with the demonstration areas and candidate non-demonstration areas arranged in columns. We generated frequency distributions when possible within categories for purposes of comparison. No statistical analyses were conducted, however. We reviewed the results with the statistician on the research team and other team members. The data were presented to TMA for consideration. Based on the criteria determined to be of primary importance, TMA selected the following catchment areas as the non-demonstration comparison areas:

- Ft. Hood as a comparison area for Ft. Carson: When compared with the other Army catchment areas we examined, Ft. Hood had the closest percentage of visits to counselors, a sufficient number beneficiaries who sought mental health service (for survey sampling), a similar-sized MTF, and the greatest number of counselors enrolled in the network.
- Wright-Patterson as a comparison area for Offutt AFB: When compared with the other Air Force catchment areas we examined, Wright-Patterson had the closest per-

centage of visits to counselors, a sufficient number of beneficiaries who sought mental health services (for survey sampling), and a similar-sized MTF.

- Luke AFB as a comparison area for the USAF Academy: When compared with the other Air Force catchment areas we examined, Luke AFB had a percentage of visits to counselors that was similar to that of the USAF Academy, a sufficient number of beneficiaries who sought mental health services (for survey sampling), and a similar-sized MTF. Luke AFB also is within Region 7/8, allowing for a within-health-care-service-region/managed care support contractor comparison.

Task 2

Obtain and analyze administrative claims (e.g., Health Care Service Record [HCSR] and Pharmacy Data Transaction Service [PDTS]) data on utilization and reimbursement for mental health services provided to covered beneficiaries within the demonstration areas, compared with utilization and reimbursement rates for similar services in non-demonstration areas (comparison areas)

Objective: To evaluate the impact of independent reimbursement of mental health services provided by licensed or certified mental health counselors on the utilization and reimbursement of such services for covered beneficiaries under the TRICARE program.

Specifically, this task was to provide (in response to the FY01 NDAA legislation):

1. A description of the extent to which expenditures for reimbursement of LMHCs change as a result of allowing the independent practice of such counselors
2. Data on utilization and reimbursement regarding non-physician mental health professionals other than LMHCs under the TRICARE program
3. Data on utilization and reimbursement regarding physicians who make referrals to, and supervise, mental health counselors
4. For each of the categories described in items 1 through 3, a comparison of data for a one-year period for the areas in which the demonstration project is being implemented with corresponding data for similar areas in which the demonstration project is not being implemented.

Task Design and Procedures: To assess the extent to which independent reimbursement of LMHCs impacts service utilization, expenditures, and treatment process outcomes, RAND conducted analyses of service claims for covered beneficiaries receiving services from mental health providers. RAND compared data on claims for care provided within the demonstration areas with claims data from non-demonstration areas (the control areas), using both one year of data pre-demonstration implementation and one year of data post-implementation. RAND also examined and compared treatment process outcomes for beneficiaries receiving mental health services from LMHCs and compared such outcomes with outcomes for beneficiaries seeking services from other mental health providers (e.g., physicians, clinical psychologists, clinical social workers). For the majority of these analyses, RAND employed a pre-post intervention evaluation methodology.

Data Sources: To examine utilization, expenditures, and treatment process outcomes, our study relied on several DoD health care data sets. We requested CYs 2002 and

2003 Health Care Service Records and pharmacy records from the Pharmacy Data Transaction Service for TRICARE beneficiaries who received mental health services in the specified catchment areas (demonstration and comparison). We also requested data from DEERS (e.g., the most recent available PITE) so that we could estimate mental health service utilization rates among eligible beneficiaries for each catchment area of interest. For mental health service users (based on the HCSR and PDTS), we also requested data from the Standard Ambulatory Data Record and the Standard Inpatient Data Record to capture any information on use of mental health services within the direct care system.

Analytic Plan:

Initial Data Extraction, Processing, and Management: We worked closely with DoD to specify the data sources, define the records and variables to be extracted, and ensure the best extraction of data for the purposes of this study. We submitted detailed data requests and a formal data-use agreement to DoD to request all health care service records/claims for mental health service users during the one-year pre-demonstration and one-year post-demonstration periods (restricted to users of mental health services provided in the specified catchment areas). The pre-demonstration and post-demonstration periods were defined using the same months (to control for any seasonal variations in mental health service utilization). To ensure comprehensiveness in our sample, we employed a broad definition of mental health service users to include beneficiaries who received TRICARE covered care, during the one-year period before the implementation of the demonstration or during the one-year period following the implementation of the demonstration, and who met one or more of the following criteria:

- Received TRICARE covered care from a mental health specialty provider (defined by the provider codes for LMHC, clinical social worker, psychologist, family/marital therapist, or psychiatrist)
- Received TRICARE covered care for a mental health service (defined by the CPT code or ICD-procedural codes for psychotherapy, psychoanalysis, psychiatric management, counseling, group/family therapy, or other care)
- Received a TRICARE covered psychotropic medication (defined by the National Drug Codes: antidepressants, stimulants, antipsychotics, anxiolytics, and other medications)
- Received TRICARE covered care in which a mental health diagnosis (ICD 9-CM codes: 292-312, 314) appeared in one of the diagnosis fields. Beneficiaries with a secondary or tertiary mental health diagnosis were considered mental health service users only if one of the other criteria was met.

Main Evaluation Analyses: After the data were formatted and prepared for analyses, using the pre-post intervention design, we examined utilization patterns and reimbursement data for a one-year period prior to the demonstration (i.e., baseline) and a one-year period of data following full implementation of the demonstration. The main evaluation analyses measured changes pre- and post-demonstration in the amount, type, and cost of mental health services provided to TRICARE beneficiaries. All analyses examined group differences between beneficiaries in the demonstration areas and those receiving care in the non-demonstration (comparison) areas by type of provider (see Chapter Five).

Analytic Questions: Our analyses were aimed at assessing the following research questions:

1. What is the impact of independent practice authority for licensed or certified mental health counselors on the expenditures for mental health services? For each question, we assessed pre- and post-demonstration changes in

   - aggregate overall expenditures (DoD and patient)
   - aggregate expenditures per provider group
   - expenditures per user.

2. What is the impact of independent practice authority for licensed or certified mental health counselors on the utilization of mental health services? We assessed changes by provider group, pre- and post-demonstration, in

   - aggregate volume of use of outpatient mental health services (number of users and number of visits)
   - the type of mental health service use (use and rate of outpatient service; use and rate of inpatient psychiatric hospitalizations)
   - the intensity of mental health service use (visits per user; combinations of services—psychotherapy alone, medication alone, psychotherapy and medication)
   - the clinical characteristics of mental health users (distribution of patients by major diagnostic category).

3. What is the impact of independent practice authority for licensed or certified mental health counselors on the utilization of health care services in general for mental health users? For each question, we assessed changes pre- and post-demonstration in

   - aggregate volume of use of outpatient and inpatient health care services among mental health users (number of visits, number of admissions, total expenditures, rate of visits, rate of admissions, and other measures of use)
   - the mix of general health care service use among mental health users.

4. What is the impact of independent practice authority for licensed or certified mental health counselors on payments for mental health services provided by mental health providers? For each area, assessed pre- and post-demonstration changes in

   - aggregate overall payments for mental health services
   - aggregate payment per provider group
   - payments per user.

Definition of Measures: Using the variables available in the administrative claim records provided by TMA, we constructed several measures of interest: outpatient visit counts, inpatient episodes, expenditures for outpatient visits and inpatient episodes, and payments to providers (see Table B.1).

Outpatient Visit Counts: We defined an outpatient visit as a "mental health" visit if the visit was to a mental health provider, there was a mental health procedure listed on the record, or there was a mental health diagnosis listed on the record. To count outpatient visits to each provider type for each user, we summed the "visits" variable across all records for the provider type (e.g., LMHC, OMH provider, psychiatrist). We did not allow for more than one visit per day to a given provider type; therefore, if a record had the same "begin" and "end" date, we capped the number of visits for that record at one. We also did not count any outpatient records that occurred during an inpatient hospitalization.

Inpatient Episodes: To identify and count inpatient episodes, we considered any HCSR non-institutional record with an "inpatient" type of service as part of an inpatient episode. Because many records labeled "inpatient" had the same "begin" and "end" dates, we strung together all inpatient events within three days of each other into the same episode. We then rolled HCSR institutional records with an overlapping date range into the same episode. Finally, we defined an inpatient episode as a "mental health episode" if there were any mental health procedures, provider types, or diagnoses for any of the records that made up an inpatient episode.

Expenditures for Outpatient Visits: Because multiple procedures and visits were often recorded on a single record with a single "amount paid" variable, we could not assign outpatient costs to a specific outpatient event. Instead, we summed costs for each individual across all mental health records and used this sum to calculate the mean outpatient expenditures per mental health service user. Likewise, to calculate total outpatient spending, we summed costs across all of a person's outpatient visits.

Expenditures for Inpatient Episodes: We calculated expenditures for an inpatient episode by summing the "amount paid" variable across all the records that made up that episode. The mean per user was calculated by dividing the sum of that variable by the total number of mental health service users or dividing it by the total number of mental health service users who had at least one inpatient mental health episode (since not all users had an inpatient episode).

Table B.1
Summary of Measures for Evaluating the Impact on Utilization and Costs of Mental Health Care Services

Measure	Description
Utilization	Visits to mental health service providers (overall volume, mean number of visits, and rates)
	Number of mental health service users (overall number and as percentage of eligible beneficiaries)
	Health care service visits for mental health service users (volume and mean)
	Type and frequency of mental health service use among users; rate of inpatient psychiatric hospitalization among mental health service users
Payments to providers	Payments (by government) made for health care services for beneficiaries receiving mental health service (total and per-user estimates)
	Payments (by government) made to providers of mental health services (total and per-user estimates)
Total expenditures	Total expenditures (amount paid by government) for health care services for beneficiaries receiving mental health services (total cost and per-user cost to government; total cost and per-user cost to patients)
	Total expenditures (amount paid by government) for services provided by mental health providers (total and per-user estimates)

Payments to Providers: To calculate payments made to the various mental health care provider groups, we totaled the "amount paid" variable, by provider type, across all outpatient visit records. We did not include records with an "outpatient" type of service that occurred during an inpatient stay.

Statistical Tests: All analyses were conducted using SAS version 8.02. To measure differences pre- and post-demonstration, where appropriate to the variable, we used chi-square for frequency distributions and tested differences in means with t-tests. To control for these population differences, we used propensity score weighting to adjust the non-demonstration group population for differences in age, sex, member category, and interactions between these characteristics. Ideally, we would have liked to conduct a multivariable analysis, using these propensity score weights, to determine the effect of the demonstration on utilization and costs. However, the data available did not provide additional variables that would be useful in predicting health care costs. In particular, we would have preferred to control for diagnoses, but diagnoses are available only on the claims data from which we determine utilization, and therefore are endogenous. Therefore, having used propensity score weights to control for variation in the only personal information we had available between the populations, we were advised by our statistical consultant to compare weighted means across the two groups. We first compared utilization across the two eligible populations, including the rate of any mental health care use and counselor use. We then compared rates of use among those seeing an LMHC. To determine if the demonstration had a significant impact on the variables of interest, we used a difference-in-difference approach to determine whether the differences (e.g., in utilization or costs) between pre- and post-demonstration in the demonstration areas are significantly different from the pre-demonstration and post-demonstration differences in the non-demonstration areas.

Task 3
Collect and analyze data on the clinical and treatment characteristics and treatment outcomes of covered beneficiaries who receive mental health services under the TRICARE program to assess the impact of independent reimbursement on health outcomes of covered beneficiaries

Objective: Evaluate the effects of the DoD demonstration of expanded access to LMHCs under TRICARE on beneficiaries' mental health processes and outcomes.

We examined differences between beneficiaries receiving mental health services in demonstration areas versus those receiving mental health services in comparison areas, and we compared beneficiaries receiving mental health services from different types of providers approximately six months after the demonstration implementation. Specifically, we aimed to do the following:

1. Describe the demographic and health characteristics of respondents compared with non-respondents (using administrative data)
2. Identify factors associated with access to care for mental health problems—e.g., reasons for seeking care, intentions to receive, care, and barriers to obtaining needed care (including a perceived social stigma)

3. Understand factors associated with adherence (and non-adherence) to treatment among those receiving mental health care services (e.g., taking psychotropic medications as recommended and completing an adequate number of psychotherapy visits)
4. Assess reported satisfaction with mental health care received from a specific provider (including communication with clinicians, information about treatment options, and patient involvement in treatment decisionmaking)
5. Evaluate mental health outcomes (diagnosis, symptom severity, and mental health functioning).

Task Design and Procedures: We used a post-demonstration mail survey of TRICARE beneficiaries to evaluate the effects of the demonstration on outcomes. The survey contained approximately 75 items, with four to five completed per minute, for a 15- to 20-minute completion time. We collected cross-sectional survey data approximately six to nine months after the full implementation of the demonstration. This data collection allowed for group comparisons to determine whether beneficiaries receive better care as a function of being in the demonstration or as a function of provider type. For example, using available HCSR data from Task 2, we examined, described, and compared characteristics of health care service use across the four beneficiary groups of interest. Beneficiary groups are those receiving services from:

1. LMHCs under the demonstration
2. LMHCs in a non-demonstration area
3. physicians (including psychiatrists and primary care physicians rendering either a defined mental health service or a service to a beneficiary with a mental health diagnosis)
4. other non-physician mental health providers.

To the greatest extent possible, procedural outcome variables were defined and assessed; those variables included rates of mental health service use, rates of overall health care service use, frequency/intensity of mental health service use, frequency/intensity of overall health care service use, and rates of use of inpatient psychiatric services. We examined and compared the clinical and treatment complexity across the four beneficiary groups. For example, based on TRICARE pharmacy data, we assessed the incidence and prevalence of mental health diagnoses and the use of services, by provider type, relative to psychotropic medication use.

Sample Selection: Based on TRICARE's assumptions on the number of beneficiaries who used mental health services during the prior month, we estimated that at least 1,200 target beneficiaries would be needed to ensure a final sample of 600 completed surveys (assuming a 50-percent response rate) for a cross-sectional survey. Because our goal was to evaluate the effect of increased access to mental health services in demonstration and non-demonstration areas and for different types of providers, we were interested in knowing whether those needing services were actually seeking care for their personal or emotional problems at the time the demonstration began. As noted earlier, we defined mental health service users broadly to include TRICARE beneficiaries with either a diagnosis of a mental disorder, a visit for a mental health service from either a specialist or a generalist, or a pharmacy claim for a psychotropic medication during the past year.

We used administrative data on mental health visits and diagnoses (at the individual-user level) to draw the sample of beneficiaries. To allow for adequate power in making comparisons across the four key comparison groups—mental health counselors in a demonstration area, mental health counselors in a non-demonstration area, other mental health specialists balanced across demonstration and non-demonstration areas, and physicians balanced across demonstration and non-demonstration areas—we sampled equal numbers of beneficiaries from each of these groups. Table B.2 shows the estimated final sample sizes and accompanying sampling probabilities.

Analytic Precision: Preliminary sample-size calculations suggested that, with this design, we would achieve more than adequate statistical power (more than 80 percent of the sample) to detect a 20-percent difference in groups (demonstration versus non-demonstration) with the proposed sample size. Power would be lower if beneficiary scores were more dispersed. However, even if the effect size was much smaller than 20 percent, there would be adequate power for analyzing differences between the two groups but the power would be low for analyzing differences by provider type.

Survey Content and Analysis: An overview of the survey content and a discussion of the analysis related to this task can be found in Chapter Two.

Task 4
Conduct relevant policy and qualitative analyses

Objective: This task was devoted to producing relevant policy and qualitative analyses to (1) describe administrative costs incurred from the requirement of documentation of referral and supervision of LMHCs, (2) assess the impact of independent reimbursement on patient confidentiality and on the willingness of LMHCs to participate in TRICARE, and (3) summarize policy requests and recommendations regarding LMHCs from plans within TRICARE.

Task Design and Procedures: Most of the actual work related to this task involved semi-structured interviewing. The first wave of interviews was conducted shortly after implementation of the demonstration (January 2003 to February 2003), and a reduced set of

Table B.2
Estimated Sample Sizes Based on Sampling Probabilities

Provider Type	Demonstration Areas	Non-Demonstration Areas	Total
Mental health counselor	150 (.25)	150 (.25)	300 (.50)
Other mental health specialist	75 (.125)	75 (.125)	150 (.25)
Physician	75 (.125)	75 (.125)	150 (.25)
Psychiatrist	38 (.0625)	37 (.0625)	75 (.125)
General medical physician	38 (.0625)	37 (.0625)	75 (.125)
Total	300 (.50)	300 (.50)	600 (1.00)

NOTE: Sampling probabilities appear in parentheses.

follow-up interviews was conducted approximately nine months after the demonstration was in place (July 2003 to September 2003). Our target interviewees included the following:

- LMHCs, psychologists, and physicians (psychiatrists and primary care providers) in the demonstration and non-demonstration areas
- representatives from the four MCSCs that provide behavioral health services under TRICARE in the regions covering the demonstration and non-demonstration areas
- a military representative(s) from the Department of Defense Mental Health Policy Group
- Congressional staffers responsible for TRICARE-related legislation
- representatives from the ACA and AMHCA.

Plans were modified to include some supplemental interviewing, depending on our ongoing results. Most of the interviews were conducted by telephone, and all participants were informed of the purpose of the interview.

Prior to conducting the interviews, we developed an interview protocol for each group of interviewees. As a product of the interviews, we produced a (typed) listing of the questions and answers from each interview. In addition, for each category of interviewee (e.g., MCSCs, LMHCs, psychiatrists), we produced a short, concise document that described the trends in responses across individual interview subjects.

In addition to the interviews, we undertook three other (smaller) non-interview tasks for the qualitative analyses. First, in assessing the impact of the intervention on confidentiality, we searched the literature to identify relevant regulatory authorities, guidance documents, and articles providing empirical evidence that might ground our discussion and analysis of confidentiality issues. Second, to the extent that our interviews pointed us toward any recent legislative proposals regarding TRICARE coverage policies (for LMHCs), we aimed to briefly examine those proposals. Finally, we compared the number of LMHCs contracted with TRICARE pre-demonstration with the number contracted one year later post-demonstration to assess the impact of the intervention on LMHCs' willingness to participate in TRICARE.

Documentation and Analysis of Qualitative Data: Interviewers input the interview responses using standard word-processing programs within 24 hours of an interview to ensure accurate recall. The study's research assistant entered the data into (tabular) files that can be electronically searched to identify themes in responses across interviewees. The data were organized into tables by interviewee type (e.g., LMHCs, MCSC), with questions listed in rows and the respondents listed in columns. This technique, supplemented by computer-based text searching, supported the identification of response trends in the qualitative data. Additionally, we created a second-order table of trends, with key themes from the interviews listed in rows, and the provider types or demonstration areas listed in columns. This tabular summary of the qualitative data facilitated the analysis of themes across the various types of interviewees. Ultimately, the results of the qualitative analysis (including the data tables) were incorporated into a narrative discussion in the final project report.

Protection of Human Subjects

This evaluation project involved the collection and analyses of primary survey data and secondary administrative data at the individual-respondent level on the use of mental health services and the collection and analyses of data obtained through interviews with individuals in official capacities related to TRICARE. All analyses were performed using de-identified data. All study procedures and protocols were reviewed and approved by the RAND Human Subjects Protection Committee to ensure efforts were taken to minimize risk of identification associated with study participation (reference file number 0152-02-03). In addition, the Department of Defense sought review and approval of the demonstration implementation procedures as an exempt human subject use study under the provisions of 45 CFR 46.101 (b) (5) from the Institutional Review Board of the Uniformed Services University for the Health Sciences (reference file number HU72FE).

The evaluation methods and the beneficiary survey instrument used to gather data directly from beneficiaries, titled "Survey of Mental Health Care Experiences," were reviewed and approved by the Defense Manpower Data Center (reference file number RCS DD-HA (OT) 2165, expiration date August 28, 2006).

Access and use of the administrative claims data were granted under a Data Use Agreement with the TRICARE Management Activity Privacy Office (reference file number DUA 0098).

RAND created and implemented an appropriate data-safeguarding and data-monitoring plan to protect and monitor data safety throughout the course of the project. We provided a copy of this plan to the TRICARE Management Activity, and it is also kept on file with the RAND Data Safeguarding Officer within the Human Subjects Protection Committee.

Beneficiary Survey: Background and Survey Instrument

This appendix provides information on the development of the TRICARE beneficiary survey questionnaire and survey sample, survey fielding activities, and the survey response. It also includes a copy of the survey questionnaire cover letters and survey instrument that were sent to potential respondents.

Questionnaire Development

Questionnaire development for the survey of TRICARE beneficiaries conducted for this study began in March 2003 with the identification of domains that would be examined in the survey. Questionnaire items were adapted from the following instruments: Experience of Care and Health Outcomes Survey—Managed Behavioral Healthcare Organization v3.0 (ECHO), Brief Patient Health Questionnaire (PHQ), and Partners in Care Brief Health Questionnaire (PIC). Most of the items focused on treatment and health status.

The title of the questionnaire was "Survey of Mental Health Care Experiences." The questionnaire was designed to elicit information from the respondent regarding his or her experiences utilizing mental health care services and coverage, recent and current health status including mental health symptoms, and attitudes about mental illness and mental health care. Demographic and other personal information (e.g. family situation, exposure to Iraq War) was also collected. The questionnaire was divided into eight sections, as listed in Table C.1.

A pilot test of the questionnaire was conducted in late May and early June 2003. It consisted of nine pilot testers filling out a self-administered questionnaire and participating in a one-on-one phone interview to provide feedback on the clarity of some of the phrasing and terminology used in the questionnaire and to explore how they thought through their answers to some key questions.

The first section of the survey questionnaire ("Treatment for Personal or Emotional Problems") was designed to assess whether the respondent had received mental health care services during the study period. Those who indicated not having received such services (which were described as including medication or other types of treatment) were instructed to skip all items related to mental health treatment.

On average, pilot testers took 20 minutes to complete the questionnaire, with only one person indicating that it took longer than expected. Overall, pilot testers found the questionnaire easy to complete. Regarding the format and appearance of the questionnaire,

Table C.1
Questionnaire Sections

Section	Description
"Treatment for Personal or Emotional Problems"	Lists examples of circumstances that might lead a person to receive mental health care services and asks respondents to indicate whether they have received these services in the past six months
"Your Counseling or Treatment"	First set of questions captures information regarding mode of care and from whom the respondent sought care; second set of questions asks about the respondent's experience receiving that care
"Your Medications and Other Health Remedies"	Items to capture information about medications used by the respondent for mental health–related ailments.
"Your Health Plan and Your Mental Health Benefits"	Questions regarding the respondent's experience with mental health care coverage
"Your Health Status"	Includes some general health items but is mostly aimed at capturing information about the respondent's mental health status
"Attitudes About Health and Health Care"	Designed to measure the respondent's perception regarding the impact of having a mental health problem and concerns regarding receiving mental health care treatment
"TRICARE Demonstration Project for Expanded Access to Mental Health Counselors"	Two items meant to assess the respondent's knowledge of the demonstration program
"About You"	Items to capture information on age, gender, education level, race/ethnicity, family situation, work status, and exposure to war in Iraq

various changes were made in response to the pilot testers' comments. For example, one pilot-test respondent indicated that the color of the cover should be soft green or blue (not bright yellow as it was) because the respondent thought that more calming and soothing colors would be important for a survey on mental health.

Based on pilot testers' input on specific items, wording changes were made to the introductory statements, question stems (a "stem" is an introductory segment of a question that is carried over to other questions), and response categories of various items. For example, regarding the list of possible reasons for obtaining mental health care, pilot testers pointed out that there was overlap between personal and family problems. It was therefore suggested that a distinction be made among family, work, and other types of personal problems.

Other refinements to the instrument's language, skip patterns, and the order of some items within sections were made prior to the main data collection. Some of these revisions, including the wording of the ethnicity/race items and drug/alcohol use items, were based on input from Defense Manpower Data Centers.

Study Sample

Table C.2 summarizes the makeup of the study sample by provider type and catchment area. Three of the catchment areas (32, 33, and 78) participated in the demonstration being evaluated, while the other three (9, 95, and 110) did not and were used for comparison purposes only. Provider type was defined based on TRICARE administrative claims data (see Chapter Five).

Table C.2
Study Sample by Catchment Area and Provider Type

Catchment Area	Provider-Type Group[a]				
	LMHC	OMH	PYSCH	Other MD	Total
Demonstration Area	307	150	75	75	607
Catchment 32	103	51	25	25	204
Catchment 33	99	51	25	25	200
Catchment 78	105	48	25	25	203
Non-Demonstration Area	293	150	75	75	593
Catchment 9	80	51	25	25	181
Catchment 95	84	51	25	25	185
Catchment 110	129	48	25	25	227
Total	600	300	150	150	1,200

[a]LMHC = potential respondent received services from a licensed or certified mental health counselor; OMH = potential respondent received services from a psychologist and/or social worker, but *not* from a licensed or certified mental health counselor; PSYCH = potential respondent received services from a psychiatrist *only;* Other MD = potential respondent received services from a non-psychiatric physician *only.*

A comparison of mailing addresses found duplicate households among 47 individuals, with 22 pairs of individuals in the same household and one set of three individuals in the same household. We kept all these individuals in the study sample.

Fielding Activities

Data collection began on September 16, 2003, and ended on February 27, 2004. Fielding procedures included three mailings of the questionnaire packet, one reminder letter mailing, and reminder phone prompts. Table C.3 outlines the fielding activities and the dates of those activities, and includes estimates of the response (completed questionnaires, or "completes") per fielding task and as a percentage of total completes received. The questionnaire packet included a cover letter on RAND Corporation letterhead, a hardcopy of the questionnaire, and a postage-paid return envelope. The first mailing also included an endorsement letter on TRICARE Management Activity letterhead signed by the Director of Health Program Analysis and Evaluation, the study sponsor. Samples of the study-packet letter and reminder letter can be found later in this appendix.

Table C.3
Fielding Activities, Sample Size, and Response

Fielding Task	Sample Size	Dates	Estimated Response per Mailing (% of Sample Size)	% of Total Response from All Fielding Activities (n = 553)
First mailing	1,200	9/16/03–9/17/03	176 (15%)	32%
Reminder letter	1,024[a]	9/22/03–9/24/03	148 (14%)	27%
Second mailing/phone prompts	764	10/23/03–11/12/03	182 (24%)	33%
Third mailing	577	1/6/04	47 (8%)	8%

[a]A reminder letter was sent to all individuals in the sample; this number excludes individuals for whom a completed survey was received prior to the date when the reminder letter was mailed (n = 176).

Phone prompts to non-respondents (approximately 844 cases) were conducted from mid-October through early November 2003. On average, those individuals received two calls during that time period. In the majority of those cases, callers from the RAND Survey Research Group (SRG) were able to leave a message for the potential respondent or talk to the potential respondent directly. Potential respondents without phone numbers or with wrong phone numbers were tracked through directory assistance.

A protocol was developed to address situations in which a respondent may express the desire to hurt himself or someone else. This desire could be expressed either in writing on the questionnaire (all questionnaires were reviewed within 24 hours of having been received by SRG) or during a phone conversation with an SRG caller. In either event, the case would be immediately referred to the appropriate TRICARE emergency assistance number in the respondent's catchment area. No incidents of a possible life-threatening situation arose during the phone prompts or were evident in returned questionnaires.

Table C.4 provides a breakdown of survey participation by sampled catchment area and provider type. Provider information for each respondent was based on TRICARE claim records (see Chapter Five for more information).

The undeliverable rate was 7 percent of the overall sample, with fairly equal distribution across the two study groups (demonstration and non-demonstration). Those cases for which the packet was returned as undeliverable by the U.S. Postal Service without any forwarding information were not tracked any further. One questionnaire packet was returned as undelivered after the end of the data collection period; that case was not included in the Undelivered column in Table C.4.

Table C.4
Final Survey Fielding Status and Response Rate

Catchment Areas and Provider Type[a]	Number Sampled	Completed Surveys Returned	Deceased	Out of Area[b]	Refused[c]	Questionnaire Packet Undelivered	Response Rate[d]
Demonstration Areas	607	271	2	0	37	40	45%
LMHC	307	137	0	0	17	21	45%
OMH	150	65	0	0	13	7	43%
PSYCH	75	41	1	0	2	6	55%
Other MD	75	28	1	0	5	6	38%
Non-Demonstration Areas	593	282	6	2	11	38	48%
LMHC	293	125	0	2	4	21	43%
OMH	150	80	1	0	6	8	54%
PSYCH	75	40	2	0	0	6	55%
Other MD	75	37	3	0	1	3	51%
Total	1,200	553	8	2	48	78	46%

[a]LMHC = potential respondent received services from a licensed or certified mental health counselor; OMH = potential respondent received services from a psychologist and/or social worker, but *not* from a licensed or certified mental health counselor; PSYCH = potential respondent received services from a psychiatrist *only;* Other MD = potential respondent received services from a non-psychiatric physician *only.*

[b]Individuals not currently living in the United States (i.e., new address provided by U.S. Postal Service was an APO [Army and Air Force Post Office] address).

[c]Includes individuals who were too sick to participate, too busy to participate, not interested, or concerned about privacy.

[d]Number of completes divided by eligible sample, where eligible sample excludes "Deceased" and "Out of Area."

Active refusals (i.e., recipients who returned a blank survey with or without an explanation or who indicated during phone prompts that they would not complete the survey) were more common among the demonstration group (6 percent) than the non-demonstration group (2 percent). Nine of the 48 refusal cases responded that they were "too old" or "too sick" to complete the survey, while the remainder said that they were not interested, too busy, or had privacy concerns.

Two of the questionnaires were returned completed, but the unique identifier on the packet label had been removed by the respondent; as such, those two questionnaires were not included in the data file of survey responses. Two respondents returned a completed questionnaire twice; in those cases, the last questionnaire received was not included in the data file of survey responses. Three completed questionnaires (all from the third mailing) were returned after the end of the data collection period; those were not included in the Completed Surveys column in Table C.4.

Data Editing and Entry

Data editing was done prior to data entry in accordance with the specifications outlined by the RAND principal investigators for this study. Data entry of close-ended survey items was completed in early March 2004 in accordance with the specifications, which included double data entry (100 percent key verification). For each case sampled, several (cumulative) ASCII data files with the close-ended survey responses, an Excel file with the "please specify" written responses, and an Excel file listing the final response status and other relevant (but not personally identified) information (e.g. provider type, catchment area) were delivered to the research team in March 2004.

OFFICE OF THE ASSISTANT SECRETARY OF DEFENSE
HEALTH AFFAIRS
SKYLINE FIVE, SUITE 810, 5111 LEESBURG PIKE
FALLS CHURCH, VIRGINIA 22041-3206

September 8, 2003

Dear Military Health System Beneficiary:

The Department of Defense (DoD) is currently conducting a demonstration program designed to allow TRICARE beneficiaries to receive care from qualifying mental health counselors without a referral or supervision from a physician. We have asked RAND, an independent non-profit research organization with offices in Arlington, Virginia, to evaluate the success of this program. To do this, RAND researchers need to collect information directly from a representative sample of military health beneficiaries, regardless of whether or not they are aware or are participating in this demonstration program.

I am writing to you today to encourage you to participate in RAND's evaluation efforts. **It is important for you to participate because your experiences and opinions matter!** Learning more about the health care needs of military health beneficiaries, and about their experiences getting the care they need, will help DoD and Congress improve benefits for all military families.

Thank you in advance for your participation in this very important effort!

RICHARD D. GUERIN, Ph.D.
Director
Health Program Analysis and Evaluation

September 2003

«fname» «mname» «lname» «suffix»
«addr1»
«city», «state» «zip»-«zipext»

Dear «fname» «lname»:

We are writing to ask you to participate in a timely study on the mental health care available to military active duty, retirees and their families. Increased stresses related to the global war on terrorism, including recent deployments to Iraq and Afghanistan, have made it extremely important for the Department of Defense to ensure that all of its beneficiaries are receiving the finest mental health care available. RAND, an independent non-profit research organization with a national reputation for quality health care research, is conducting this study on behalf of the Department of Defense TRICARE. The main purpose of this study is to gain a better understanding from TRICARE beneficiaries of issues regarding access to and satisfaction with mental health care services.

Enclosed you will find a questionnaire and a postage-paid envelope. We would appreciate it if you could complete the questionnaire and return it in the postage-paid envelope as soon as possible.

You were selected to receive this questionnaire from a list obtained from the Department of Defense of military health beneficiaries in your catchment area. *As an individual eligible to receive health care benefits from the military, your experiences with and views regarding your health care coverage are extremely valuable to the Department of Defense and Congress as they work on improving the health care benefits offered to military personnel and their families.*

While your input is extremely valuable, your participation in this study is voluntary. If you choose not to participate, it will not affect the benefits that you and your family personally receive. Please be assured that RAND will keep your responses strictly confidential, and that RAND will not release any information that can be linked to you, unless RAND is required by law to do so. Furthermore, RAND will not give to anyone at TRICARE, including health care providers or plan administrators, or anyone affiliated with the military your individual answers to the survey. By returning your completed questionnaire, you agree to participate in this study and for your responses to be combined with administrative information obtained from TRICARE about health care services used by beneficiaries in your catchment area. However, RAND will not have access to your personal medical health records and your responses to this survey will never be linked to individual medical information.

If you have any questions about the study or have trouble completing the questionnaire, please call Ana Suárez toll free at 1-888-345-6377.

We look forward to your input on this very important issue.

Lisa S. Meredith, Ph.D.
Co-Principal Investigator

Terri Tanielian, M.A.
Co-Principal Investigator

Survey of Mental Health Care Experiences

Center for Military Health Policy Research
and
National Defense Research Institute

RAND
1200 South Hayes Street
Arlington, Virginia 22202-5050

© RAND 2003

INSTRUCTIONS FOR COMPLETING THIS QUESTIONNAIRE

About this questionnaire

This questionnaire was designed as part of a larger study being conducted by RAND on behalf of the   Department of Defense. The information being collected in this questionnaire will help the Department of Defense and TRICARE better understand how to improve mental health care coverage for all military health beneficiaries.

How to fill out this questionnaire

- Answer all the questions by checking the box to the left of your answer, unless otherwise indicated.

- If after checking an answer you then decide you want to change it, simply cross out the answer you want to change and check your new answer.

- You are sometimes told to skip over some questions in this survey. When this happens you will see an arrow with a note that tells you what question to answer next, like this:

    ₁☑ Yes   ➡ **If Yes, Go to Question 3**
    ₂☐ No

- If your response is other than those specifically listed, you are asked to include more information in the line provided as follows:

    ☐ Other *(please specify):* _____

- In one question, you are asked to write your response in the blank space provided.

Returning the questionnaire

We have included a pre-addressed, postage-paid envelope for you to return the completed questionnaire directly to RAND. If you have any questions, please call Ana Suárez, RAND Survey Coordinator, toll free at 888-345-6377.

TREATMENT FOR PERSONAL OR EMOTIONAL PROBLEMS

PEOPLE CAN GET COUNSELING, TREATMENT OR MEDICATION FOR MANY DIFFERENT REASONS, SUCH AS FOR:

- FEELING DEPRESSED, ANXIOUS, OR "STRESSED OUT"

- WORK PROBLEMS (LIKE WHEN ONE IS HAVING DIFFICULTIES GETTING ALONG WITH PEOPLE AT WORK)

- FAMILY PROBLEMS (LIKE MARRIAGE PROBLEMS OR WHEN PARENTS AND CHILDREN HAVE TROUBLE GETTING ALONG)

- OTHER PERSONAL PROBLEMS (LIKE WHEN A LOVED ONE DIES OR WHEN ONE IS HAVING DIFFICULTIES GETTING ALONG WITH FRIENDS)

- NEEDING HELP WITH DRUG OR ALCOHOL USE

- FOR MENTAL OR EMOTIONAL ILLNESS

1. **In the last 6 months, did you personally get <u>counseling, treatment or medication</u> for any of these reasons?**

   1☐ Yes ➡ *If Yes, Go to Question 2*
   2☐ No ➡ *If No, Go to Question 50 on Page 10*

YOUR COUNSELING OR TREATMENT

THIS SECTION ASKS ABOUT <u>YOUR</u> EXPERIENCES WI COUNSELING OR TREATMENT IN THE <u>LAST 6 MONTH</u> WHEN ANSWERING THESE QUESTIONS, INCLUDE A COUNSELING AND TREATMENT RECEIVED DURIN OUTPATIENT VISITS WITH ANY MENTAL HEALTH CA PROVIDER OR FOR ANY MENTAL HEALTH REASON. D <u>NOT</u> INCLUDE COUNSELING OR TREATMENT RECEIVE DURING AN OVERNIGHT STAY IN A HOSPITAL OR FRO A SELF-HELP GROUP.

2. **In the last 6 months, did you try to get <u>professional counseling on the phone</u> for yourself?**

   1☐ Yes
   2☐ No ➡ *If No, Go to Question 4*

3. **In the last 6 months, how often did you <u>get</u> the professional counseling you needed <u>on the phone</u>?**

   1☐ Never
   2☐ Sometimes
   3☐ Usually
   4☐ Always

4. **At any time in the last 6 months, did you need counseling or treatment <u>immediately</u> because of an emergency or crisis?**

   1☐ Yes
   2☐ No ➡ *If No, Go to Question 6*

5. **In the last 6 months, when you needed counseling or treatment <u>immediately</u> for an emergency or crisis, how often did you see someone as soon as you wanted?**

   1☐ Never
   2☐ Sometimes
   3☐ Usually
   4☐ Always

6. In the last 6 months (not counting times you needed counseling or treatment immediately for an emergency or crisis), did you make any appointments for counseling or treatment?

₁☐ Yes
₂☐ No   ➡ *If No, Go to Question 8*

7. In the last 6 months (not counting times you needed counseling or treatment immediately for an emergency or crisis), how often did you get an appointment for counseling or treatment as soon as you wanted?

₁☐ Never
₂☐ Sometimes
₃☐ Usually
₄☐ Always

8. In the last 6 months (not counting times you needed counseling or treatment immediately for an emergency or crisis), what kinds of mental health care providers did you talk to or see for counseling or treatment? *You can check more than one.*

₁☐ Psychiatrist
₂☐ Psychologist
₃☐ Social worker
₄☐ Psychiatric nurse
₅☐ Mental health counselor
₆☐ Family physician or other primary health care provider
₇☐ Other – please specify:

_____

₈☐ Don't know/Can't remember

9. Of the mental health providers you saw for counseling or treatment during the past 6 months (not counting times you needed counseling or treatment immediately for an emergency or crisis):

a. Who did you see most recently?
Please check one only.

₁☐ Psychiatrist
₂☐ Psychologist
₃☐ Social worker
₄☐ Psychiatric nurse
₅☐ Mental health counselor
₆☐ Family physician or other primary health care provider
₇☐ Other – please specify:

_____

₈☐ Don't know/Can't remember

b. How did you first find out about the provider you saw most recently?
Please check one only.

₁☐ Through your health plan's toll free telephone line
₂☐ From your health plan's provider directory
₃☐ Recommended by another provider
₄☐ Recommended by a friend
₅☐ Other – please specify:

_____

_____

10. In the last 6 months (not counting emergency rooms or crisis centers), how many times did you go to an office, clinic, or other treatment program to get counseling, treatment, or medication for yourself?

₀☐ None   ➡ *If None, Go to Question 12*
₁☐ 1 time only
₂☐ 2 to 10 times
₃☐ 11 to 20 times
₄☐ 21 times or more

11. In the last 6 months, how often were you seen **within 15 minutes** of your appointment time?

   1☐ Never
   2☐ Sometimes
   3☐ Usually
   4☐ Always

THE REMAINING QUESTIONS IN THIS SECTION ARE ABOUT ALL THE COUNSELING OR TREATMENT YOU GOT IN THE LAST 6 MONTHS DURING OFFICE, CLINIC, EMERGENCY ROOM, AND CRISIS CENTER VISITS AS WELL AS OVER THE PHONE. *PLEASE DO THE BEST YOU CAN TO INCLUDE ALL THE DIFFERENT PEOPLE YOU WENT TO FOR COUNSELING OR TREATMENT IN YOUR ANSWERS.*

12. In the last 6 months, how often did the people you went to for counseling or treatment **listen carefully to you**?

   1☐ Never
   2☐ Sometimes
   3☐ Usually
   4☐ Always

13. In the last 6 months, how often did the people you went to for counseling or treatment **explain things** in a way you could understand?

   1☐ Never
   2☐ Sometimes
   3☐ Usually
   4☐ Always

14. In the last 6 months, how often did the people you went to for counseling or treatment **show respect for what you had to say**?

   1☐ Never
   2☐ Sometimes
   3☐ Usually
   4☐ Always

15. In the last 6 months, how often did th people you went to for counseling or treatment **spend enough time** with you?

   1☐ Never
   2☐ Sometimes
   3☐ Usually
   4☐ Always

16. In the last 6 months, how often did you **feel safe** when you were with the people you went to for counseling or treatment?

   1☐ Never
   2☐ Sometimes
   3☐ Usually
   4☐ Always

17. In the last 6 months, how often were you **involved as much as you wanted** your counseling or treatment?

   1☐ Never
   2☐ Sometimes
   3☐ Usually
   4☐ Always

18. In the last 6 months, did anyone talk you about **whether to include** your family or friends in your counseling c treatment?

   1☐ Yes
   2☐ No

19. In the last 6 months, were you given information about **different kinds** of counseling or treatment that are available?

   1☐ Yes
   2☐ No

20. In the last 6 months, did you take any prescription medications as part of your treatment for personal or emotional problems?

₁☐ Yes
₂☐ No ➡ *If No, Go to Question 22*

21. In the last 6 months, were you told what side effects of those medications to watch for?

₁☐ Yes
₂☐ No

22. In the last 6 months, were you told about self-help or support groups, such as consumer-run groups or 12-step programs?

₁☐ Yes
₂☐ No

23. In the last 6 months, were you given as much information as you wanted about what you could do to manage your condition?

₁☐ Yes
₂☐ No

24. In the last 6 months, were you given information about your rights as a patient?

₁☐ Yes
₂☐ No

25. In the last 6 months, did you feel you could refuse a specific type of medication or treatment?

₁☐ Yes
₂☐ No

26. In the last 6 months, as far as you know did anyone you went to for counseling or treatment share information with others that should have been kept private?

₁☐ Yes
₂☐ No

27. Does your language, race, religion, ethnic background, or culture make any difference in the kind of counseling or treatment you need?

₁☐ Yes
₂☐ No ➡ *If No, Go to Question 29*

28. In the last 6 months, was the care you received responsive to those needs listed in Question 27 above?

₁☐ Yes
₂☐ No

29. Using any number from 0 to 10, where 0 is the worst counseling or treatment possible and 10 is the best counseling or treatment possible, what number would you use to rate all your counseling or treatment in the last 6 months?

☐ 0   Worst counseling or treatment possible
☐ 1
☐ 2
☐ 3
☐ 4
☐ 5
☐ 6
☐ 7
☐ 8
☐ 9
☐ 10   Best counseling or treatment possible

4

30. In the last 6 months, how much were you helped by the counseling or treatment you got?

- $_1$☐ Not at all
- $_2$☐ A little
- $_3$☐ Somewhat
- $_4$☐ A lot

31. Overall, how dissatisfied or satisfied were you with the health care available to you for personal or emotional problems in the last 6 months?

- $_1$☐ Very dissatisfied
- $_2$☐ Somewhat dissatisfied
- $_3$☐ Neither satisfied nor dissatisfied
- $_4$☐ Somewhat satisfied
- $_5$☐ Very satisfied

32. In the last 6 months, was there any time when you didn't get as much mental health care for emotional or personal problems as you needed, or had delays in getting care?

- $_1$☐ Yes
- $_2$☐ No ➡ *If No, Go to Question 34*

33. What was <u>the main reason</u> you didn't get as much help as you needed or had delays in getting care in the last 6 months? *Please check one only.*

- $_1$☐ I was worried about how much it would cost
- $_2$☐ I was worried about what others might think
- $_3$☐ I had difficulties getting a referral from my Military Treatment Facility provider
- $_4$☐ I had difficulties finding a provider or making an appointment
- $_5$☐ I had scheduling problems because of other personal responsibilities such as home, family or work
- $_6$☐ Other reason – please specify:

_____

_____

_____

_____

Continue ▨

4. How often was *each* of the following statements true for you during the <u>past 4 weeks</u>?	None of the time	A little of the time	Some of the time	A good bit of the time	Most of the time	All of the time
a. I had a hard time doing what my mental health care provider(s) suggested I do.............	1☐	2☐	3☐	4☐	5☐	6☐
b. I followed the suggestions of my mental health care provider(s) exactly...............	1☐	2☐	3☐	4☐	5☐	6☐
c. I was unable to do what was necessary to follow the treatment plans proposed by my mental health care provider(s)...............	1☐	2☐	3☐	4☐	5☐	6☐
d. I found it easy to do the things my mental health care provider(s) suggested I do.............	1☐	2☐	3☐	4☐	5☐	6☐
e. Overall, I was able to do what my mental health care provider(s) told me...............	1☐	2☐	3☐	4☐	5☐	6☐

35. During the <u>past 4 weeks</u>, did you receive counseling (for example, talk therapy) from a mental health care provider?

1☐ Yes
2☐ No ➡ *If No, Go to Question 37*

36. How often was *each* of the following statements true for you during the <u>past 4 weeks</u>?	None of the time	A little of the time	Some of the time	A good bit of the time	Most of the time	All of the time
a. I showed up to all my therapy or counseling sessions................	1☐	2☐	3☐	4☐	5☐	6☐
b. I avoided situations that trigger my symptoms................	1☐	2☐	3☐	4☐	5☐	6☐
c. I tried to play an active role in my therapy or counseling................	1☐	2☐	3☐	4☐	5☐	6☐

37. At any time during the <u>past 6 months</u>, did you take any prescription, nonprescription or over-the-counter medications because you were feeling depressed, stressed out or anxious, or because you were experiencing difficulty sleeping, low energy or pain?

  ₁☐ Yes
  ₂☐ No ➡ *If No, Go to Question 42*

38. In the spaces below, please list up to 4 prescription, nonprescription or over-the-counter medications you took in the <u>last 6 months</u> because you were feeling depressed, stressed out or anxious, or because you were experiencing difficulty sleeping, low energy or pain. Also, please indicate how many days in total you took each medication and whether you are still taking the medication:

a. Name of medication (enter one name in each space below)	b. Total # of days you took this medication in the last 6 months:	c. Are you still taking this medication?	d. Why did you stop taking this medication? *You can check more than one*
1.	₁☐ 2 weeks or less ₂☐ 3 to 4 weeks ₃☐ more than 1 month but less than 3 months ₄☐ 3 months or more	₁☐ Yes ₂☐ No ➡	₁☐ You were having side effects ₂☐ You felt worse or NO better ₃☐ You felt better ₄☐ You feared becoming addicted ₅☐ It cost too much ₆☐ It was too hard to take ₇☐ You didn't need it ₈☐ Some other reason – please specify
2.	₁☐ 2 weeks or less ₂☐ 3 to 4 weeks ₃☐ more than 1 month but less than 3 months ₄☐ 3 months or more	₁☐ Yes ₂☐ No ➡	₁☐ You were having side effects ₂☐ You felt worse or NO better ₃☐ You felt better ₄☐ You feared becoming addicted ₅☐ It cost too much ₆☐ It was too hard to take ₇☐ You didn't need it ₈☐ Some other reason – please specify
3.	₁☐ 2 weeks or less ₂☐ 3 to 4 weeks ₃☐ more than 1 month but less than 3 months ₄☐ 3 months or more	₁☐ Yes ₂☐ No ➡	₁☐ You were having side effects ₂☐ You felt worse or NO better ₃☐ You felt better ₄☐ You feared becoming addicted ₅☐ It cost too much ₆☐ It was too hard to take ₇☐ You didn't need it ₈☐ Some other reason – please specify
4.	₁☐ 2 weeks or less ₂☐ 3 to 4 weeks ₃☐ more than 1 month but less than 3 months ₄☐ 3 months or more	₁☐ Yes ₂☐ No ➡	₁☐ You were having side effects ₂☐ You felt worse or NO better ₃☐ You felt better ₄☐ You feared becoming addicted ₅☐ It cost too much ₆☐ It was too hard to take ₇☐ You didn't need it ₈☐ Some other reason – please specify

9. How often was *each* of the following statements true for you during the <u>past 6 months</u>?	None of the time	A little of the time	Some of the time	A good bit of the time	Most of the time	All of the time
a. I took my medications for the recommended length of time.........	1☐	2☐	3☐	4☐	5☐	6☐
b. I took the correct dosage for my medications............................	1☐	2☐	3☐	4☐	5☐	6☐
c. I skipped taking my medications............................	1☐	2☐	3☐	4☐	5☐	6☐
d. I delayed getting refills for my medications............................	1☐	2☐	3☐	4☐	5☐	6☐

40. Hypericum, also known as St. Johns' Wort, is an herbal substance that can be purchased without a prescription. Have you used hypericum in the <u>last 6 months</u> because you were feeling depressed, stressed out or anxious, or because you were experiencing difficulty sleeping or low energy?

1☐ Yes
2☐ No   ➡   *If No, Go to Question 42*

41. How often have you used hypericum in the <u>last 6 months</u>?

1☐ 2 weeks or less
2☐ 3 to 4 weeks
3☐ more than 1 month but less than 3 months
4☐ 3 months or more

42. **Which health plan did you use for <u>all or most</u> of your mental health counseling or treatment in the <u>last 6 months</u>?**
Please check one only

- 1☐ TRICARE Prime
- 2☐ TRICARE Senior Prime or TRICARE Plus
- 3☐ TRICARE Extra or Standard (CHAMPUS)
- 4☐ TRICARE for Life (Medicare plus TRICARE)
- 5☐ Other health insurance (please specify: _____
- 6☐ I didn't use any health plan; I paid for it out of pocket most of the time

43. **In the last 6 months, did <u>you use up all your benefits</u> for counseling or treatment?**

- 1☐ Yes
- 2☐ No ➡ *If No, Go to Question 46*

44. **At the time benefits were used up, did you think you <u>still needed</u> counseling or treatment?**

- 1☐ Yes
- 2☐ No ➡ *If No, Go to Question 46*

45. **Were you told about <u>other ways</u> to get counseling, treatment or medicine?**

- 1☐ Yes
- 2☐ No

46. **In the last 6 months, did you call <u>customer service</u> to get information or help about counseling or treatment?**

- 1☐ Yes
- 2☐ No ➡ *If No, Go to Question 48*

47. **In the last 6 months, how much of a problem, if any, was it to <u>get the help you needed</u> when you called customer service?**

- 1☐ A big problem
- 2☐ A small problem
- 3☐ Not a problem

48. **In the last 6 months, did you need to get approval to receive any counseling or treatment?**

- 1☐ Yes
- 2☐ No ➡ *If No, Go to Question 50*

49. **In the last 6 months, how much of a problem, if any, were <u>delays</u> in counseling or treatment while you wait for approval?**

- 1☐ A big problem
- 2☐ A small problem
- 3☐ Not a problem

Continue ➡

50. During the last 4 weeks, how much have you been bothered by any of the following problems?

	Not bothered	Bothered a little	Bothered a lot
a. Stomach pain	1 ☐	2 ☐	3 ☐
b. Back pain	1 ☐	2 ☐	3 ☐
c. Pain in your arms, legs, or joints (knees, hips, etc.)	1 ☐	2 ☐	3 ☐
d. Menstrual cramps or other problems with your periods	1 ☐	2 ☐	3 ☐
e. Pain or problems during sexual intercourse	1 ☐	2 ☐	3 ☐
f. Headaches	1 ☐	2 ☐	3 ☐
g. Chest pain	1 ☐	2 ☐	3 ☐
h. Dizziness	1 ☐	2 ☐	3 ☐
i. Fainting spells	1 ☐	2 ☐	3 ☐
j. Feeling your heart pound or race	1 ☐	2 ☐	3 ☐
k. Shortness of breath	1 ☐	2 ☐	3 ☐
l. Constipation, loose bowels, or diarrhea	1 ☐	2 ☐	3 ☐
m. Nausea, gas, or indigestion	1 ☐	2 ☐	3 ☐

51. In general, how would you rate your overall health now?

1 ☐ Excellent
2 ☐ Very good
3 ☐ Good
4 ☐ Fair
5 ☐ Poor

52. During the last 2 weeks, how often have you been bothered by any of the following problems?

	Not at all	Several days	More than half the days	Nearly every day
a. Little interest or pleasure in doing things.............................	1 ☐	2 ☐	3 ☐	4 ☐
b. Feeling down, depressed, or hopeless...........................	1 ☐	2 ☐	3 ☐	4 ☐
c. Trouble falling or staying asleep, or sleeping too much.......	1 ☐	2 ☐	3 ☐	4 ☐
d. Feeling tired or having little energy..............................	1 ☐	2 ☐	3 ☐	4 ☐
e. Poor appetite or overeating...................................	1 ☐	2 ☐	3 ☐	4 ☐
f. Feeling bad about yourself — or that you are a failure or have let yourself or your family down............................	1 ☐	2 ☐	3 ☐	4 ☐
g. Trouble concentrating on things, such as reading the newspaper or watching television................................	1 ☐	2 ☐	3 ☐	4 ☐
h. Moving or speaking so slowly that other people could have noticed? Or the opposite — being so fidgety or restless that you have been moving around a lot more than usual.....	1 ☐	2 ☐	3 ☐	4 ☐
i. Thoughts that you would be better off dead or of hurting yourself in some way...	1 ☐	2 ☐	3 ☐	4 ☐

53. Have you ever had an anxiety attack – suddenly feeling fear or panic?

1 ☐ Yes
2 ☐ No ➡ *If No, Go to Question 58*

54. Have some of these attacks come suddenly out of the blue — that is, in situations where you don't expect to be nervous or uncomfortable?

1 ☐ Yes
2 ☐ No

55. Do these attacks bother you a lot or are you worried about having another attack?

1 ☐ Yes
2 ☐ No

56. In the last 4 weeks, have you had an anxiety attack?

1 ☐ Yes
2 ☐ No

57. Think about the last time you had a bad anxiety attack:	YES	NO	Not Sure
a. Were you short of breath?	1☐	2☐	3☐
b. Did your heart race, pound, or skip?	1☐	2☐	3☐
c. Did you have chest pain or pressure?	1☐	2☐	3☐
d. Did you sweat?	1☐	2☐	3☐
e. Did you feel as if you were choking?	1☐	2☐	3☐
f. Did you have hot flashes or chills?	1☐	2☐	3☐
g. Did you have nausea or an upset stomach, or the feeling that you were going to have diarrhea?	1☐	2☐	3☐
h. Did you feel dizzy, unsteady, or faint?	1☐	2☐	3☐
i. Did you have tingling or numbness in parts of your body?	1☐	2☐	3☐
j. Did you tremble or shake?	1☐	2☐	3☐
k. Were you afraid you were dying?	1☐	2☐	3☐

58. Over the last 4 weeks, how often have you been bothered by any of the following problems?

	Not at all	Several days	More than half the days
a. Feeling nervous, anxious, on edge, or worrying a lot about different things	1☐	2☐	3☐

If you checked "Not at all" to Question 58a, go to Question 59.

	Not at all	Several days	More than half the days
b. Feeling restless so that it is hard to sit still	1☐	2☐	3☐
c. Getting tired very easily	1☐	2☐	3☐
d. Muscle tension, aches, or soreness	1☐	2☐	3☐
e. Trouble falling asleep or staying asleep	1☐	2☐	3☐
f. Trouble concentrating on things, such as reading a book or watching TV	1☐	2☐	3☐
g. Becoming easily annoyed or irritable	1☐	2☐	3☐

59. In the last 6 months, have you had any emotional or personal problems that have made it difficult for you to do your work, take care of things at home, or get along with other people?

₁☐ None
₂☐ Yes, somewhat difficult
₃☐ Yes, very difficult
₄☐ Yes, extremely difficult

60. In general, how would you rate your overall mental health now?

₁☐ Excellent
₂☐ Very good
₃☐ Good
₄☐ Fair
₅☐ Poor

61. Compared to 6 months ago, how would you rate your ability to deal with daily problems now?

₁☐ Much better
₂☐ A little better
₃☐ About the same
₄☐ A little worse
₅☐ Much worse

62. Compared to 6 months ago, how would you rate your ability to deal with social situations now?

₁☐ Much better
₂☐ A little better
₃☐ About the same
₄☐ A little worse
₅☐ Much worse

63. Compared to 6 months ago, how would you rate your ability to accomplish the things you want to do now?

₁☐ Much better
₂☐ A little better
₃☐ About the same
₄☐ A little worse
₅☐ Much worse

64. Compared to 6 months ago, how would you rate your problems or symptoms now?

₁☐ Much better
₂☐ A little better
₃☐ About the same
₄☐ A little worse
₅☐ Much worse

Continue ➡

5. If you were applying for a job, how much difficulty do you think you would have getting the job if the employer thought you had a recent history of the following:	A lot of difficulty	Some difficulty	A little difficulty	No difficulty	Not sure
a. Diabetes...	1 ☐	2 ☐	3 ☐	4 ☐	5 ☐
b. High blood pressure............................	1 ☐	2 ☐	3 ☐	4 ☐	5 ☐
c. HIV or AIDS......................................	1 ☐	2 ☐	3 ☐	4 ☐	5 ☐
d. Mental health problems e.g. depression or anxiety...	1 ☐	2 ☐	3 ☐	4 ☐	5 ☐
e. Visiting a mental health provider..............	1 ☐	2 ☐	3 ☐	4 ☐	5 ☐

6. Please indicate how strongly you agree or disagree with the following statements:	Strongly Agree	Somewhat Agree	Neither Agree nor Disagree	Somewhat Disagree	Strongly Disagree
a. In order to get a job a person with mental health problems will have to hide his or her mental health history.............	1 ☐	2 ☐	3 ☐	4 ☐	5 ☐
b. There is no reason for a person to hide the fact that he or she has a history of mental health problems....................	1 ☐	2 ☐	3 ☐	4 ☐	5 ☐
c. If a person has a serious mental illness, the best thing to do is keep it a secret....	1 ☐	2 ☐	3 ☐	4 ☐	5 ☐
d. If I had a close relative who had been treated for a serious mental illness, I would advise him or her not to tell anyone about it......................................	1 ☐	2 ☐	3 ☐	4 ☐	5 ☐
e. I rarely feel the need to hide the fact that I received mental health treatment.........	1 ☐	2 ☐	3 ☐	4 ☐	5 ☐

14

67. Think about a future time when you might need or want care for emotional or personal problems. Please indicate how likely or unlikely it is that you <u>might not get</u> the care you need or want because of the following <u>reasons</u>:	Very likely	Somewhat likely	Neither likely nor unlikely	Somewhat unlikely	Very unlikely
a. I would worry about the cost..................	1 ☐	2 ☐	3 ☐	4 ☐	5 ☐
b. I would worry about the effect on my own or a family member's military career.........	1 ☐	2 ☐	3 ☐	4 ☐	5 ☐
c. I would not be able to get a referral from my Military Treatment Facility provider......	1 ☐	2 ☐	3 ☐	4 ☐	5 ☐
d. The mental health provider does not accept my health insurance.............................	1 ☐	2 ☐	3 ☐	4 ☐	5 ☐
e. My health plan does not pay for the type of treatment I would need.........................	1 ☐	2 ☐	3 ☐	4 ☐	5 ☐
f. I would not be able to find out where to go for help.................................	1 ☐	2 ☐	3 ☐	4 ☐	5 ☐
g. I would not be able to get to the mental health provider's office when it is open......	1 ☐	2 ☐	3 ☐	4 ☐	5 ☐
h. The mental health provider's office is too far from my house or work.....................	1 ☐	2 ☐	3 ☐	4 ☐	5 ☐
i. I have difficulties getting through to the mental health provider's office on the telephone to make an appointment...........	1 ☐	2 ☐	3 ☐	4 ☐	5 ☐
j. I do not think I could be helped...............	1 ☐	2 ☐	3 ☐	4 ☐	5 ☐
k. I would be embarrassed to discuss my problem with anyone...........................	1 ☐	2 ☐	3 ☐	4 ☐	5 ☐
l. I would be afraid of what others would think	1 ☐	2 ☐	3 ☐	4 ☐	5 ☐
m. I would be afraid of losing pay from work	1 ☐	2 ☐	3 ☐	4 ☐	5 ☐
n. I would need someone to take care of my children, elderly parents or disabled spouse...	1 ☐	2 ☐	3 ☐	4 ☐	5 ☐

TRICARE DEMONSTRATION PROJECT FOR EXPANDED ACCESS TO MENTAL HEALTH COUNSELORS

UNDER THE TRICARE DEMONSTRATION PROJECT FOR EXPANDED ACCESS TO MENTAL HEALTH COUNSELORS, LICENSED AND CERTIFIED MENTAL HEALTH COUNSELORS CAN NOW PROVIDE SERVICES TO COVERED TRICARE BENEFICIARIES WITHOUT A REFERRAL FROM A PHYSICIAN AND WITHOUT HAVING TO BE SUPERVISED BY A PHYSICIAN.

68. Before receiving this questionnaire, did you know about this TRICARE demonstration project?

₁ ☐ Yes
₂ ☐ No ➡ *If No, Go to Question 70*

69. How did you learn about this TRICARE demonstration project? *You can check more than one.*

₁ ☐ Discussed it with a mental health provider
₂ ☐ Read an article about the demonstration in a DOD/TRICARE newsletter
₃ ☐ Heard about the demonstration from family or friends
₄ ☐ Other (please specify):

_____

ABOUT YOU

70. What is your age now?

₁ ☐ 18 to 24
₂ ☐ 25 to 34
₃ ☐ 35 to 44
₄ ☐ 45 to 54
₅ ☐ 55 to 64
₆ ☐ 65 to 74
₇ ☐ 75 or Older

71. Are you male or female?

₁ ☐ Male
₂ ☐ Female

72. What is the highest grade or level of school that you have <u>completed</u>?

₁ ☐ 8th grade or less
₂ ☐ Some high school, but did not graduate
₃ ☐ High school graduate or GED
₄ ☐ Some college or 2-year degree
₅ ☐ 4-year college graduate
₆ ☐ Post-graduate degree

73. Are you Spanish/Hispanic/Latino?

₁ ☐ No, not Spanish/Hispanic/Latino
₂ ☐ Yes, Mexican, Mexican-American, Chicano, Puerto Rican, Cuban or other Spanish/Hispanic/Latino

74. What is your race?
You can check more than one to indicate what you consider yourself.

₁ ☐ White
₂ ☐ Black or African-American
₃ ☐ American Indian or Alaska Native
₄ ☐ Asian (e.g. Asian Indian, Chinese, Filipino, Japanese, Korean, Vietnamese)
₅ ☐ Native Hawaiian or other Pacific Islander (e.g. Samoan, Guamanian or Chamorro)

75. Were you born in the United States?

₁ ☐ Yes ➡ *If Yes, Go to Question 78*
₂ ☐ No

76. About how many years have you lived in the United States? *Your best guess is fine. If less than a year, please enter "1".*

_____ year(s)

77. How well do you <u>speak</u> English?

₁ ☐ Very well
₂ ☐ Well
₃ ☐ Not well
₄ ☐ Not at all

78. Do you have any children or stepchildren?

₁☐ Yes
₂☐ No ➡ *If No, Go to Question 81*

79. In the last 6 months, have any of your children or stepchildren received counseling, treatment or medicine for emotional or behavioral problems?

₁☐ Yes
₂☐ No ➡ *If No, Go to Question 81*

80. In the last 6 months, have any of your children or stepchildren been placed in a detention center or a residential treatment center?

₁☐ Yes
₂☐ No

81. Are you currently living alone?

₁☐ Yes ➡ *If Yes, Go to Question 84*
₂☐ No

82. Which of the following living arrangements describe your situation at this time?
You can check more than one.

₁☐ Currently living with a spouse or partner
₂☐ Currently living with your children or others who are related to you
₃☐ Currently living with other people (other than a partner) who are not related to you

83. Please indicate how strongly you agree or disagree with the following statement:

During the past 6 months I have felt very close to the people I live with.

₁☐ Strongly agree
₂☐ Somewhat agree
₃☐ Neither agree nor disagree
₄☐ Somewhat disagree
₅☐ Strongly disagree

84. Please select the item that best descri your current employment status. By fu time we mean 35 or more hours per week. *Check only one.*

₁☐ Working full-time ➡ *Go to Question* for pay
₂☐ Working part-time for pay
₃☐ Not working for pay

85. Are you currently not working full-time because of your health?

₁☐ Yes
₂☐ No

86. Are you or any member of your family enrolled in the Exceptional Family Member Program?

₁☐ Yes
₂☐ No
₃☐ Don't know/Not sure

87. Were any of your family members or close friends deployed for the recent in Iraq?

₁☐ None ➡ *If None, Go to Question 8*
₂☐ Spouse
₃☐ Other family member (please specify): _____
₄☐ Close friend

88. Are any of them back from their tour o duty?

₁☐ None
₂☐ Spouse
₃☐ Other immediate family
₄☐ Close friend

Continue

89. In the last 6 months, have you received any counseling or treatment from a mental health care provider because of personal or emotional problems related to the recent war in Iraq?

$_1\square$ Yes
$_2\square$ No

90. Did someone help you complete this survey?

$_1\square$ Yes ➡ *If Yes, Go to Question 91*
$_2\square$ No ➡ *If No, Please return the survey in the postage-paid envelope*

91. Who helped you to complete this form?

$_1\square$ A family member
$_2\square$ A friend
$_3\square$ Someone else

****

Is there anything else you would like to share with us? Your comments are greatly appreciated.

_____

_____

_____

_____

_____

_____

_____

_____

_____

_____

_____

_____

Please return the questionnaire in the envelope provided to:

RAND
1200 South Hayes Street
Arlington, VA 22202-5050
Attn: Ana Suárez

If you have any questions, please call Ana Suárez, RAND Survey Coordinator, toll free at 888-345-6377.

THANK YOU!

Beneficiary Survey: Supplemental Data Tables

The previous appendix provides detailed information on the design and fielding of the beneficiary survey, as well as a copy of the survey itself. In this appendix, we provide tables with detailed data to illustrate results from analyses of that survey.

Table D.1 shows the scoring rules and weighted descriptive statistics for each variable derived from the beneficiary survey, with the exception of the design characteristics that were obtained from TRICARE administrative data to determine eligibility for the survey sample. The table displays the variables by type of measure.

Table D.1
Description of Variables Derived from the Beneficiary Survey

Variable	Scoring	Mean/ % (Standard Deviation)
Design Characteristics		
Demonstration catchment area	1 if demonstration catchment area, 0 otherwise	50.2
Saw a mental health care provider	1 if sampled because respondent saw a mental health provider in the past 6 months, 0 otherwise	90.0
Received a mental health procedure	1 if sampled because respondent received a mental health procedure (e.g., a CPT procedure code for psychotherapy, medication management, psychoanalysis, etc.: 90805, 90811, 90807, 90812, etc.) in the past 6 months, 0 otherwise	23.2
Had a psychiatric diagnosis	1 if sampled because respondent had a psychiatric diagnosis in the past 6 months, 0 otherwise	99.2
Study/Survey Characteristics		
Proxy responder	1 if a designated person completed the survey on the respondent's behalf, 0 otherwise	6.7
Exposure to demonstration	1 if beneficiary reported knowing about the TRICARE demonstration project before receiving this questionnaire	4.8
Demographic Characteristics		
Age Group		
18–24	1 if age 18–24, 0 otherwise	16.0
25–34	1 if age 25–34, 0 otherwise	19.1
35–44	1 if age 35–44, 0 otherwise	21.3
45–54	1 if age 45–54, 0 otherwise	20.1

Table D.1—Continued

Variable	Scoring	Mean/ % (Standard Deviation)
55–64	1 if age 55–64, 0 otherwise	13.9
65+	1 if age 65+, 0 otherwise	9.6
Male	1 if male, 0 otherwise	17.8
Education		
High school or less	1 if high school or less, 0 otherwise	24.9
Some college	1 if some college, 0 otherwise	47.9
College graduate	1 if college graduate, 0 otherwise	27.2
Latino ethnicity	1 if Latino, 0 otherwise	6.0
Race		
White	1 if white, 0 otherwise	84.7
Black	1 if black, 0 otherwise	8.7
Other	1 if other race/ethnicity, 0 otherwise	6.6
U.S.-born	1 if born in the United States, 0 otherwise	88.8
Have children	1 if have child(ren), 0 otherwise	79.9
Child(ren) received mental health care	1 if child(ren) received mental health care, 0 otherwise	24.1
Live alone	1 if live alone, 0 otherwise	12.4
Working	1 if currently working, 0 otherwise	44.9
Not working due to health problem(s)	1 if not currently working due to health problem(s), 0 otherwise	20.4
Health Characteristics		
Mental Health Symptoms and Disorder		
Somatic symptoms	1 if beneficiary meets criteria for probable somatic disorder based on the PHQ, 0 otherwise	25.9
Major depression	1 if beneficiary meets criteria for probable major depression based on the PHQ, 0 otherwise	19.7
Depression score	Count of reported frequency of PHQ depression symptoms experienced in the past 2 weeks re-scored as: 0 = not at all, 1 = several days, 3 = more than half the days, 3 = nearly every day	7.78 (910.0)
Other depression	1 if beneficiary meets criteria for probable depression other than major depressive disorder based on the PHQ, 0 otherwise	8.4
Panic disorder	1 if beneficiary meets criteria for probable panic depression based on the PHQ, 0 otherwise	45.2
Other anxiety	1 if beneficiary meets criteria for probable anxiety disorder other than panic disorder based on the PHQ, 0 otherwise	13.9
Emotional problems affecting functioning	Beneficiary reports having experienced emotional or personal problems making it difficult to function in the past 6 months; rescored as 1 if there are difficulties, 0 otherwise	68.6
Overall mental health	Rating of current overall mental health: 1 = excellent, 2 = very good, 3 = good, 4 = fair, 5 = poor (reversed so that a higher score indicates better health)	3.0 (1.5)
General Health		
Overall health	Rating of current overall health: 1 = excellent, 2 = very good, 3 = good, 4 = fair, 5 = poor (reversed so that a higher score indicates better health)	3.2 (1.5)

Table D.1—Continued

Variable	Scoring	Mean/ % (Standard Deviation)
Use of Services and Treatments		
Received mental health care	1 if received mental health care in the past 6 months, 0 otherwise	85.3
Received counseling from a mental health services provider	1 if received counseling from a mental health services provider in the past 4 weeks, 0 otherwise	50.8
Took any medication for a mental health problem	1 if took any type of medication (prescription [Rx], non-Rx, or over-the-counter) for a mental health problem in the past 6 months, 0 otherwise	75.5
Took a prescription medication for a mental health problem	1 if took a prescription medication as part of treatment for personal or emotional problems in the past 6 months, 0 otherwise	76.7
Took *Hypericum* (Saint-John's-wort) for a mental health problem	1 if took *Hypericum* for a mental health problem in the past 6 months, 0 otherwise	1.8
Used an antidepressant	Used an antidepressant for a mental health problem in the past 6 months, 0 otherwise	52.7
Used antianxiety medication	Used an antianxiety medication for a mental health problem in the past 6 months, 0 otherwise	9.1
Used an antipsychotic	Used an antipsychotic medication for a mental health problem in the past 6 months, 0 otherwise	9.8
Used a benzodiazapene	Used a benzodiazapene for a mental health problem in the past 6 months, 0 otherwise	15.3
Used a mood stabilizer	Used a mood stabilizer for a mental health problem in the past 6 months, 0 otherwise	7.2
Used a stimulant	Used a stimulant for a mental health problem in the past 6 months, 0 otherwise	2.6
Used substance abuse medication	Used a substance abuse medication for a mental health problem in the past 6 months, 0 otherwise	8.7
Used another non-mental health medication	Used another medication for a non–mental health problem in the past 6 months, 0 otherwise	21.4
Access to Mental Health Care		
Any experience with barrier to mental health care	1 if any of 6 barriers to care (Q33) was reported, 0 otherwise (among beneficiaries reporting that they did not get as much mental health care as needed in the past 6 months)	28.0
Perceived barriers to mental health care (0–14)	Count of 1 through 14 for 14 potential barriers to mental health care (Q67) if beneficiary reported any of the barriers as being "very likely" or "somewhat likely"	3.5 (4.8)
Barriers by Type		
Cost	1 if perceived barriers due to cost, 0 otherwise	56.8
Career	1 if perceived barriers due to professional concerns, 0 otherwise	38.6
Cannot be helped	1 if perceived barriers due to not thinking they could be helped, 0 otherwise	12.5

Table D.1—Continued

Variable	Scoring	Mean/ % (Standard Deviation)
Stigma	1 if perceived barriers due to social stigma, 0 otherwise	30.2
Access	1 if perceived barriers due to access, 0 otherwise	54.0
Family	1 if perceived family-related barriers, 0 otherwise	23.2
Job-related stigma (1–5)	Minimum of Q65a–e[b]	1.8 (1.8)
Need for secrecy (1–5)	Average of Q66a–e (after reversing a, c, d, and e)	3.0 (1.5) Alpha = .80
Adherence to Treatment		
General adherence (0–100)	Average of Q34a–e (after reversing a and c) and then transformed to a linear 0–100 distribution	73.8 (20.9) Alpha = .84
Medication adherence (0–100)	Average of Q39a–d (after reversing c and d) and then transformed to a linear 0–100 distribution	92.3 (13.0) Alpha = .68
Counseling adherence (0–100)	Average of Q36a–c and then transformed to a linear 0–100 distribution	13.6 Alpha = .54
HEDIS Indicators[a]		
Rated counseling and treatment 9 or 10 on 0–10 scale	0–10 scale rescored as 1 if rated treatment at high end of scale (9 or 10), 0 otherwise	47.1/ 69.8
Reported "always" got urgent treatment as soon as needed	1 if always/usually or always got urgent treatment as soon as needed, 0 otherwise	47.0/57.6
Reported "always" got appointment as soon as he or she wanted	1 if always/usually or always got appointment as soon as he or she wanted, 0 otherwise	54.1/85.2
Got help by telephone	1 if always/usually or always got help by telephone, 0 otherwise	19.9/26.6
Never waited more than 15 minutes for appointment	1 if never waited more than 15 minutes, 0 otherwise	57.1/ 86.8
Helped "a lot" by treatment	1 if a lot/somewhat or a lot of help from treatment, 0 otherwise	56.7/84.5
Clinicians listen carefully	1 if clinicians always/usually or always listen carefully, 0 otherwise	67.8/91.3
Clinicians explain things	1 if clinicians always/usually or always explain things, 0 otherwise	67.7/91.9
Clinicians show respect	1 if clinicians always/usually or always show respect, 0 otherwise	75.3/91.9
Clinicians spend enough time	1 if clinicians always/usually or always spend enough time, 0 otherwise	61.1/85.7
Feel safe with clinicians	1 if always/usually or always feel safe with clinicians, 0 otherwise	76.1/92.1
Involved as much as he or she wanted in treatment	1 if always/usually or always involved as much as he or she wanted in treatment, 0 otherwise	63.6/86.3
Deal with symptoms or problems	1 if patient rates her or his ability to deal with symptoms or problems much better/a little better or much better as compared with 6 months ago	31.9/62.8

Table D.1—Continued

Variable	Scoring	Mean/ % (Standard Deviation)
Accomplish things	1 if patient rates her or his ability to accomplish things much better/a little better or much better compared with 6 months ago	27.9/57.8
Deal with social situations	1 if patient rates her or his ability to deal with social situations much better/a little better or much better as compared with 6 months ago	33.0/59.6
Deal with daily problems	1 if patient rates her or his ability to deal with daily problems much better/a little better or much better as compared with 6 months ago	39.5/69.1
No problems with helpfulness of customer service	1 if no problem with helpfulness of customer service, 0 otherwise	62.9
Told about self-help/consumer-run programs	1 if told about self-help or consumer run programs, 0 otherwise	28.4
Told about treatment options	1 if told about different treatments that are available for condition, 0 otherwise	53.6
Told about side effects of medications	1 if told about the side effects of medications, 0 otherwise	81.4
Talked about including family and friends in treatment	1 if talked about including family and friends in treatment, 0 otherwise	57.5
Given as much information as wanted to manage condition	1 if given as much information as wanted to manage condition, 0 otherwise	75.1
Given information about rights as a patient	1 if given information about rights as a patient, 0 otherwise	82.3
Patient feels that he or she could refuse a specific type of treatment	1 if patient feels that he or she could refuse a specific type of treatment, 0 otherwise	89.2
Confident about privacy of treatment Information	1 if confident about privacy of treatment information, 0 otherwise	97.4
Care responsive to cultural needs	1 if care responsive to cultural needs, 0 otherwise	77.4
No delays in treatment while waiting for plan approval	1 if no problem with delays in treatment while waiting for plan approval, 0 otherwise	71.3
Iraq War Exposure		
Anyone close deployed	1 if a close friend or family member was deployed to the war in Iraq, 0 otherwise	31.5
Not yet back from active duty	1 if close friend or family member deployed to the war in Iraq has not returned from active duty, 0 otherwise	17.1
Received mental health care due to war	1 if reported receiving mental health care due to the war in Iraq, 0 otherwise	12.5

[a]Indicators with multiple versions separated by a slash represent different cut-offs for dichotomizing the measures. The first uses only the highest response category relative to all other categories, and the second, more liberal definition includes the top two response categories.

[b]Refers to a survey item number; see Appendix C.

Tables D.2 through D.14 display, for each variable in Table D.1, the weighted bivariate means (for continuous measures) or percentage (for binary indicators) for comparing TRICARE beneficiaries in the demonstration catchment areas with beneficiaries in the non-demonstration catchment areas. Statistical significance for these two-group comparisons is shown in the form of t-tests for continuous measures or chi-square statistics for categorical or binary measures. Tables are organized by type of measure (e.g., sample characteristics, symptoms and disorder, perceived improvement, use of services).

Table D.2
Sample Selection Characteristics

Characteristic	Non-Demonstration Areas (%) (N = 282)	Demonstration Areas (%) (N = 271)	χ^2
Saw a mental health care provider	88.9	91.0	1.44
Received a mental health procedure	18.6	27.9	14.40***
Had a psychiatric diagnosis	98.9	99.4	1.15
Had a mental health prescription	63.0	61.8	0.20

NOTES: *p < .05, **p < .01, ***p < .001. Percentages may not add to 100 due to rounding.

Table D.3
Mental Health Symptoms and Probable Disorder

Characteristic	Non-Demonstration Areas (%) (N = 282)	Demonstration Areas (%) (N = 271)	χ^2
Somatic symptoms	28.0	25.8	0.69
Major depression	19.7	20.1	0.03
Other depression	8.9	8.1	0.28
Panic disorder	43.0	47.4	2.30
Other anxiety	18.0	18.8	0.09
Emotional problems affecting functioning	66.3	72.6	5.55*

NOTES: *p < .05, **p < .01, ***p < .001. Percentages may not add to 100 due to rounding.

Table D.4
Perceived Improvement from Six Months Ago

Characteristic	Non-Demonstration Areas R1 (R2) (%) (N = 282)	Demonstration Areas R1 (R2) (%) (N = 271)	χ^2
Deal with symptoms or problems	31.0 (62.4)	32.7 (63.3)	0.41 (0.10)
Accomplish things	27.4 (55.7)	28.4 (59.8)	0.16 (2.06)
Deal with social situations	32.8 (61.2)	33.3 (58.0)	0.05 (1.22)
Deal with daily problems	36.6 (69.0)	42.4 (69.1)	4.19 (0.00)

NOTES: *p < .05, **p < .01, ***p < .001. R1 = highest response category only, R2 = top two response categories. Percentages may not add to 100 due to rounding.

Table D.5
Use of Mental Health Services and Treatments

Characteristic	Non-Demonstration Areas (%) (N=282)	Demonstration Areas (%) (N=271)	t or χ^2
Received mental health care in past six months	83.2	87.5	4.32*
Received counseling from a mental health services provider in past four weeks	54.9	46.8	6.51*
Took any medication (Rx, non-Rx, or over-the-counter) for a mental health problem in the past six months	76.1	77.4	0.324
Took an Rx medication as part of treatment for personal or emotional problems in the past six months	75.4	75.6	0.01
Took *Hypericum* (Saint-John's-wort) for a mental health problem in the past six months	0.88	2.7	3.83

NOTE: *p < .05, **p < .01, ***p < .001

Table D.6
Access to Mental Health Care

Characteristic	Non-Demonstration Areas (%) (N = 282)	Demonstration Areas (%) (N = 271)	t or χ^2
Number of barriers to mental health care (0–14) [a]	3.4	3.7	−1.04
By type			
Cost (%)	56.3	59.5	1.26
Career (%)	37.8	41.2	1.43
Cannot be helped (%)	11.1	14.7	3.23
Stigma (%)	29.8	32.2	0.81
Access (%)	54.3	56.8	0.77
Family (%)	19.4	28.6	13.23**
General adherence (scale of 0–100)	72.7	73.8	−0.54
Medication adherence (scale 0–100)	91.3	91.8	−0.31
Counseling adherence (scale 0–100)	80.9	79.8	0.59
Job-related stigma (scale 1–5)	1.8	1.8	0.42
Need for secrecy (scale 1–5)	3.0	3.0	−0.57

NOTES: *p < .05, **p < .01, ***p < .001. Percentages may not add to 100 due to rounding.
[a] 28 percent reported at least one barrier.

Table D.7
HEDIS Indicators of Access to Mental Health Care

Characteristic	Non-Demonstration Areas R1 (R2) (%) (N = 282) [a]	Demonstration Areas R1 (R2) (%) (N = 271) [a]	χ^2
Rated counseling and treatment 9 or 10 on a scale of 0–10	49.1 (66.7)	45.2 (72.7)	1.58 (4.45)
Reported "always" got urgent treatment as soon as needed	28.5 (48.2)	44.9 (66.6)	7.77** (9.32)**
Reported "always" got appointment as soon as he or she wanted	52.8 (83.2)	55.4 (87.1)	0.65 (2.76)
Got help by telephone	13.1 (16.4)	25.7 (35.4)	7.38** (13.51)***
No delays in treatment from waiting for plan approval	69.1	73.3	1.12
No problems with customer service	59.6	65.8	1.33
Helped "a lot" by treatment	56.9 (84.9)	56.4 (84.1)	0.02 (0.11)

Table D.7—Continued

Characteristic	Non-Demonstration Areas R1 (R2) (%) (N = 282)[a]	Demonstration Areas R1 (R2) (%) (N = 271)[a]	χ^2
Told about self-help/consumer-run programs	29.4	27.5	0.47
Told about different treatments available for condition	52.1	54.9	0.82
Never waited more than 15 minutes	58.5 (89.5)	55.7 (84.3)	0.80 (5.88)*

NOTES: *$p < .05$, **$p < .01$, ***$p < .001$. R1 = highest response category only, R2 = top two response categories. Percentages may not add to 100 due to rounding.
[a]Among beneficiaries who reported receiving counseling, treatment, or medication for a personal or emotional problem in the past 12 months.

Table D.8
HEDIS Indicators of Clinician Communication

Characteristic	Non-Demonstration Areas R1 (R2) (%) (N = 282)[a]	Demonstration Areas R1 (R2) (%) (N = 271)[a]	χ^2
Clinicians listen carefully	70.2 (91.0)	65.7 (91.8)	2.31 (0.23)
Clinicians explain things	69.2 (92.7)	66.2 (91.3)	1.06 (0.61)
Clinicians show respect	77.5 (91.4)	73.2 (92.3)	2.50 (0.31)
Clinicians spend enough time	60.0 (83.8)	62.1 (87.5)	0.49 (2.85)
Feel safe with clinicians	77.3 (90.0)	76.2 (94.0)	0.16 (5.54)*
Involved as much as he or she wanted in treatment	66.7 (85.1)	60.7 (87.5)	3.94* (1.23)

NOTES: *$p < .05$, **$p < .01$, ***$p < .001$. Percentages may not add to 100 due to rounding.
[a]Among beneficiaries who reported receiving counseling, treatment, or medication for a personal or emotional problem in the past 12 months.

Table D.9
HEDIS Indicators of General Communication

Characteristic	Non-Demonstration Areas (%) (N=282)[a]	Demonstration Areas (%) (N=271)[a]	χ^2
Told about side effects of medications	77.6	85.0	6.89**
Talk about including family and friends in treatment	51.5	63.2	14.26***
Given as much information as wanted to manage condition	75.3	75.0	0.01
Given information about rights as a patient	79.6	84.7	4.47*
Patient feels that he or she could refuse a specific type of treatment	88.5	89.8	0.50
Confident about privacy of treatment information	97.6	97.1	0.26
Care responsive to cultural needs	70.4	84.0	4.70*

NOTES: * $p < .05$, ** $p < .01$, ***$p < .001$. Percentages may not add to 100 due to rounding.
[a]Among beneficiaries who reported receiving counseling, treatment, or medication for a personal or emotional problem in the past 12 months.

Table D.10
Respondents Taking Psychotropic Medications, by Type of Medication

Characteristic	Non-Demonstration Areas (%) (N = 282)	Demonstration Areas (%) (N = 271)	χ^2
Antidepressant	50.7	54.8	2.0
Antianxiety	10.3	7.8	2.2
Antipsychotic	6.6	13.0	14.1***
Benzodiazapene	18.2	12.4	7.9**
Mood stabilizer	5.4	9.1	6.2*
Stimulant	1.8	3.3	2.8
Substance abuse	0.3	1.1	2.9
Other non–mental health medication	17.9	24.8	8.5**

NOTES: *p < .05, **p < .01, ***p < .001. Percentages may not add to 100 due to rounding.

Table D.11
Reasons for Discontinuing an Antidepressant Among Those No Longer Taking the Medication

Characteristic	Non-Demonstration Areas (%) (N = 282)	Demonstration Areas (%) (N = 271)	χ^2
Side effects	36.4	43.3	0.97
Felt worse	39.8	42.7	0.18
Felt better	13.2	9.4	0.72
Too hard to take	0.0	3.0	2.76
Did not need it	6.9	8.2	0.13
Stopped per doctor's orders	29.0	30.0	0.02
Stopped on own	2.4	14.0	8.36**
Safety concerns	5.5	3.0	0.79

NOTES: *p < .05, **p < .01, ***p < .001. Percentages may not add to 100 due to rounding.

Table D.12
Treatment Duration Among Those Taking an Antidepressant Medication

Characteristic	Non-Demonstration Areas (%) (N = 282)	Demonstration Areas (%) (N = 271)
2 weeks or less	1.4	4.0
3–4 weeks	3.4	0.9
> 1 month but < 3 months	10.6	12.5
3 months or more	83.5	82.6

NOTES: *p < .05, **p < .01, ***p < .001. Percentages may not add to 100 due to rounding. The chi-square test for the overall effect of duration by demonstration and non-demonstration areas is 9.3.

Table D.13
Iraq War Exposure (Weighted)

Characteristic	Non-Demonstration Areas (%) (N = 282)	Demonstration Areas (%) (N = 271)	χ^2
Anyone close deployed	28.5	34.5	5.03*
Not yet back from duty[a]	14.4	19.8	6.22*
Received MH care due to war	12.7	12.8	0.00

NOTES: *p < .05, **p < .01, ***p < .001. Percentages may not add to 100 due to rounding.
[a]Among those reporting deployment.

Tables D.14 through D.17 show the multivariable regression results for selected measures of health characteristics, perceived access to care and use of services, adherence to care, and satisfaction with care. For continuous outcomes, we show the betas from the ordinary least-squares regression runs, and for binary outcomes we present the ORs along with the 95-percent confidence intervals. All models are weighted to represent the 1,200 TRICARE beneficiaries sampled and to whom we mailed a survey packet.

Table D.14
Odds Ratios and 95-Percent Confidence Intervals for the Effects on Access to Mental Health Care

Variable	Received Mental Health Care (N = 472)	Received Counseling from Mental Health Care Provider (N = 401)	Took Any Medication for Mental Health Problem (N = 406)	Took Rx Medication for Mental Health Problem (N = 412)
Demonstration catchment area	1.17 [0.79, 1.73]	0.68 [0.51, 0.90]**	1.05 [.75, 1.47]	1.10 [0.79, 1.54]
Age 25–34	0.64 [0.30, 1.35]	1.30 [0.81, 2.08]	0.74 [.43, 1.29]	1.24 [0.72, 2.13]
Age 35–44	0.83 [0.39, 1.79]	1.86 [1.16, 2.99]**	1.22 [.70, 2.16]	1.14 [0.66, 1.96]
Age 45–54	1.17 [0.52, 2.61]	2.04 [1.27, 3.28]**	2.20 [1.19, 4.05]*	2.43 [1.33, 4.42]**
Age 55 or over	0.23 [0.11, .47]**	1.83 [1.10, 3.04]*	1.97 [1.04, 3.74]*	2.30 [1.21, 4.37]*
Male	0.45 [0.27, .74]**	0.73 [0.48, 1.09]	0.25 [0.16, .40]***	0.34 [0.21, 0.55]***
Some college education	1.26 [0.77, 2.06]	0.83 [0.58, 1.19]	1.03 [0.66, 1.62]	0.60 [0.37, 0.97]*
College graduate	1.46 [0.82, 2.59]	0.86 [0.56, 1.31]	0.61 [0.37, 1.01]	0.27 [0.16, 0.46]***
Latino	1.05 [0.39, 2.82]	0.88 [0.46, 1.66]	0.75 [0.36, 1.56]	0.69 [0.33, 1.45]
Black	0.76 [0.39, 1.54]	0.56 [0.32, 0.97]*	0.94 [0.50, 1.77]	0.61 [0.34, 1.10]
Other	2.65 [0.79, 8.84]	1.25 [0.71, 2.21]	2.16 [1.00, 4.70]	2.85 [1.24, 6.55]*
Live alone	0.75 [0.42, 1.32]	1.18 [0.76, 1.83]	0.64 [0.39, 1.07]	0.63 [0.39, 1.03]
Working	0.59 [0.39, .89]*	0.73 [0.55, 1.00]*	0.61 [0.43 .87]**	0.63 [0.44, 0.89]**
Barriers: Cost	0.52 [0.32, .85]**	0.84 [0.60, 1.18]	0.70 [0.46, 1.06]	0.85 [0.57, 1.27]
Barriers: Professional	0.88 [0.55, 1.40]	1.24 [0.90, 1.72]	1.11 [0.76, 1.63]	0.94 [0.64, 1.38]
Barriers: Cannot be helped	0.90 [0.43, 1.86]	0.64 [0.41, 1.02]	0.95 [0.54, 1.66]	0.90 [0.51, 1.58]
Barriers: Stigma	1.09 [0.66, 1.82]	1.21 [0.85, 1.71]	1.59 [1.03, 2.46]*	2.84 [1.80, 4.47]***
Barriers: Access	2.06 [1.27, 3.35]*	1.22 [0.87, 1.71]	1.35 [0.90, 2.04]	1.03 [0.69, 1.53]
Barriers: Family	1.97 [1.06, 3.65]*	0.96 [0.68, 1.36]	0.90 [0.60, 1.36]	0.64 [0.43, 0.96]*
Job-related stigma, scale of 1–5	0.81 [0.69, .94]**	0.91 [0.81, 1.04]	0.88 [0.77, 1.02]	0.93 [0.81, 1.08]
Need for secrecy, scale of 1–5	1.15 [0.92, 1.43]	1.21 [1.04, 1.42]*	1.19 [0.99, 1.43]	1.17 [0.98, 1.41]
Anyone close deployed to Iraq War	0.59 [0.37, .94]*	1.74 [1.26, 2.41]***	0.74 [0.51, 1.08]	0.58 [0.40, .084]**

NOTES: *p<.05, ** p<.01, ***p<.001. All estimates are based on weighted and adjusted logistic regression models. Percentages may not add to 100 due to rounding.

Table D.15
Beta Coefficients for the Effects on Adherence to Health Care

Variable	General Adherence (SE) (N=464)	Adherence to Mental Health Medication (SE) (N=393)	Adherence to Mental Health Counseling (SE) (N=399)
Intercept	74.09 (5.94)	87.15 (4.43)	90.00 (5.98)
Demonstration catchment area	1.89 (2.14)	0.79 (1.56)	1.09 (2.01)
Age 25–34	3.89 (3.66)	8.89 (2.72)**	−4.71 (3.52)
Age 35–44	−0.13 (3.63)	8.11 (2.62)**	−1.82 (3.49)
Age 45–54	2.23 (2.62)	7.92 (2.60)**	2.96 (3.47)
Age 55 or over	6.07 (3.92)	9.76 (2.87)***	3.47 (3.80)
Male	1.26 (3.13)	2.19 (2.44)	−2.36 (3.03)
Some college education	2.53 (2.70)	−0.64 (1.97)	−1.91 (2.50)
College graduate	5.54 (3.22)	1.32 (2.36)	3.02 (2.97)
Latino	−7.20 (5.00)	4.61 (3.49)	7.87 (4.76)
Black	4.39 (4.12)	−1.79 (3.13)	6.33 (4.14)
Other	2.40 (4.12)	−3.78 (2.95)	2.19 (3.85)
Live alone	−2.25 (3.09)	3.14 (2.55)	4.80 (3.08)
Working	0.07 (2.23)	−1.29 (1.67)	−2.71 (2.21)
Barriers: Cost	−2.12 (2.60)	−0.66 (1.88)	−0.22 (2.47)
Barriers: Professional	−2.50 (2.43)	−0.74 (1.88)	2.28 (2.33)
Barriers: Cannot be helped	−12.29 (3.63)***	1.33 (2.57)	−1.11 (3.66)
Barriers: Stigma	−0.02 (2.63)	−0.44 (1.91)	−0.19 (2.30)
Barriers: Access	−4.03 (2.64)	3.43 (1.95)	−3.72 (2.42)
Barriers: Family	−4.31 (2.68)	−3.77 (1.95)	−4.61 (2.52)
Job-related stigma, scale of 1–5	1.34 (0.98)	0.23 (0.74)	−0.40 (0.97)
Need for secrecy, scale of 1–5	−0.96 (1.17)	−1.26 (0.86)	−1.48 (1.13)
Anyone close deployed to Iraq	1.70 (2.70)	1.86 (1.99)	−0.64 (2.48)
Received mental health care due to Iraq War[a]	−1.59 (3.42)	0.40 (2.41)	1.21 (2.81)

NOTES: *p < .05, **p < .01, ***p < .001. All estimates are based on weighted and adjusted logistic regression models. SE = stand error. Percentages may not add to 100 due to rounding.
[a] Among those reporting deployment.

Table D.16
Odds Ratios and 95-Percent Confidence Intervals for the Effects on Mental Health Status

Variable	Emotional or Personal Problems Affected Functioning (N = 474)	Probable Major Depression (N = 474)	Probable Panic Disorder (N = 475)	Probable Somatic Disorder (N = 460)
Demonstration catchment area	1.34 [1.00, 1.81]	0.92 [.64, 1.30]	1.04 [0.98, 1.37]	0.87 [0.64, 1.19]
Age 25–34	0.83 [0.48, 1.41]	1.19 [.64, 2.21]	.83 [0.52, 1.34]	0.98 [0.58, 1.66]
Age 35–44	0.78 [.047, 1.29]	2.72 [1.47, 5.03]**	0.76 [0.48, 1.22]	2.05 [1.23, 3.42]**
Age 45–54	2.19 [1.26, 3.83]**	3.25 [1.77, 5.96]***	0.78 [0.49, 1.24]	2.15 [1.28, 3.61]**
Age 55 or over	1.00 [.59, 1.69]	2.91 [1.52, 5.58]**	0.36 [0.22, 0.60]***	1.41 [0.80, 2.48]
Male	0.92 [0.62, 1.38]	1.19 [.73, 1.94]	0.56 [0.38, 0.85]**	0.71 [0.44, 1.14]
Some college education	0.92 [0.63, 1.35]	0.95 [0.62, 1.45]	0.89 [0.63, 1.25]	0.90 [0.62, 1.31]
College graduate	0.59 [0.38, .91]*	0.34 [0.20, 0.59]***	0.46 [0.30, 0.70]***	0.32 [0.20, 0.52]***
Latino	3.00 [1.29, 6.99]*	0.79 [0.33, 1.93]	1.93 [1.02, 3.67]*	0.29 [0.10, 0.80]*
Black	0.24 [.14, .42]***	0.42 [0.21, 0.84]*	0.82 [.49, 1.39]	0.90 [0.51, 1.57]
Other	1.19 [0.64, 2.21]	1.24 [0.64, 2.38]	1.39 [0.79, 2.42]	1.75 [0.99, 3.12]
Live alone	1.19 [0.74, 1.91]	1.12 [0.65, 1.94]	0.96 [0.62, 1.50]	1.07 [0.66, 1.73]
Working	0.84 [0.62, 1.15]	0.89 [0.61, 1.28]	0.58 [0.43, 0.77]***	0.89 [0.64, 1.23]
Barriers: Cost	0.71 [0.49, 1.03]	1.03 [0.67, 1.58]	0.78 [0.56, 1.11]	1.15 [0.77, 1.68]
Barriers: Professional	1.23 [0.87, 1.74]	1.25 [0.85, 1.84]	1.47 [1.06, 2.02]*	0.77 [0.54, 1.11]
Barriers: Cannot be helped	0.91 [0.54, 1.52]	3.43 [2.11, 5.58]***	0.74 [0.47, 1.16]	2.04 [1.27, 3.25]**
Barriers: Stigma	1.31 [0.89, 1.93]	1.01 [0.67, 1.52]	0.92 [0.65, 1.30]	1.45 [0.99, 2.10]
Barriers: Access	1.47 [1.03, 2.11]*	2.00 [1.22, 2.91]**	1.34 [0.96, 1.88]	1.60 [1.10, 2.35]*
Barriers: Family	1.99 [1.31, 3.01]**	1.81 [1.19, 2.75]**	1.24 [0.88, 1.76]	1.44 [0.98, 2.12]
Job-related stigma, scale of 1–5	1.05 [0.92, 1.19]	0.75 [0.62, 0.91]**	0.90 [0.80, 1.02]	0.81 [0.70, 0.94]**
Need for secrecy, scale of 1–5	1.49 [1.27, 1.77]***	1.34 [1.10, 1.63]**	1.42 [1.22, 1.66]***	1.04 [0.87, 1.24]
Anyone close deployed to Iraq	1.42 [0.97, 2.07]	2.13 [1.38, 3.27]***	1.01 [0.71, 1.43]	1.22 [0.83, 1.79]
Received mental health care due to Iraq War[a]	5.01 [2.46, 10.17]***	1.30 [0.76, 2.22]	3.89 [2.36, 6.39]***	2.75 [1.71, 4.42]***

NOTES: *p < .05, ** p < .01, ***p < .001. All estimates are based on weighted and adjusted logistic regression models. Percentages may not add to 100 due to rounding.
[a]Among those reporting deployment.

Table D.17
Odds Ratios for the Effects on Satisfaction with Mental Health Care Services[a]

Variable	Overall Rating of Counseling/Treatment (N = 399)	Got Urgent Treatment as Soon as Needed (N = 103)[b]	Got Appointment as Soon as Wanted (N = 361)	Got Help by Telephone (N = 109)[b]	Never Waited More Than 15 Minutes to See Clinician (N = 392)
Demonstration catchment area	1.95 [1.40, 2.70]***	3.97 [1.76, 8.95]***	1.54 [0.96, 2.50]	3.59 [1.59, 8.12]**	0.54 [0.34, 0.86]*
Age 25–34	1.39 [0.85, 2.28]	8.41 [2.10, 33.60]**	0.62 [0.32, 1.22]	3.92 [0.65, 23.85]	1.84 [0.94, 3.58]
Age 35–44	1.65 [1.00, 2.73]	51.05 [9.41, 276.99]***	2.08 [0.95, 4.58]	10.09 [1.95, 52.06]**	1.88 [0.92, 3.85]
Age 45–54	3.32 [1.93, 5.72]***	7.99 [2.06, 31.04]**	1.62 [0.78, 3.37]	10.95 [2.17 55.35]**	1.91 [0.93, 3.94]
Age 55 or over	2.23 [1.25, 3.98]**	18.78 [4.24, 83.12]***	3.12 [1.15, 8.43]*	10.83 [2.07, 56.55]**	1.74 [0.72, 4.19]
Male	0.79 [0.48, 1.30]	1.66 [0.59, 4.71]	0.82 [0.40, 1.71]	0.06 [0.01, 0.34]**	1.05 [0.50, 2.22]
Some college education	1.01 [0.67, 1.52]	0.17 [0.06, 0.49]***	1.66 [0.93, 2.96]	0.95 [0.37, 2.46]	1.03 [0.60, 1.80]
College graduate	1.04 [0.63, 1.70]	0.13 [0.04, 0.45]**	1.40 [0.67, 2.95]	1.51 [0.52, 4.38]	0.94 [0.46, 1.92]
Latino	0.57 [0.28, 1.18]	1.13 [0.28, 4.53]	0.89 [0.31, 2.57]	1.54 [0.32, 7.36]	0.33 [0.14, 0.77]**
Black	1.17 [0.63, 2.19]	1.44 [0.48, 4.31]	0.97 [0.37, 2.57]	< 0.00 [<0.0, >0999]	1.34 [0.54, 3.33]
Other	0.39 [0.22, .69]**	1.19 [0.35, 4.09]	1.43 [0.58, 3.55]	1.01 [0.27, 3.80]	0.66 [0.30, 1.43]
Live alone	0.51 [0.32, .83]**	0.22 [0.07, .72]*	0.50 [0.26, 0.99]*	0.38 [0.12, 1.23]	1.03 [0.49, 2.15]
Working	0.54 [0.39, .76]***	0.74 [0.33, 1.65]	0.74 [0.45, 1.21]	0.97 [0.41, 2.28]	0.92 [0.58, 1.48]
Barriers: Cost	0.93 [0.63, 1.39]	1.41 [0.56, 3.58]	2.23 [1.24, 4.02]**	3.25 [1.28, 8.27]*	1.18 [0.69 2.02]
Barriers: Professional	0.84 [0.59, 1.20]	3.27 [1.37, 7.82]**	0.60 [0.35, 1.02]	1.13 [0.47, 2.73]	0.61 [0.36, 1.02]
Barriers: Cannot be helped	0.78 [0.48, 1.25]	0.24 [0.07, 0.76]*	0.34 [0.19, .61]***	0.79 [0.25, 2.51]	0.59 [0.31, 1.10]
Barriers: Stigma	1.10 [0.75, 1.63]	1.84 [0.76, 4.47]	0.53 [0.31, 0.91]*	0.93 [0.33, 2.58]	0.85 [0.49, 1.47]
Barriers: Access	0.45 [0.30, .67]***	0.50 [0.19, 1.37]	0.26 [0.13, 0.50]***	0.81 [0.33, 2.01]	1.52 [0.88, 2.64]
Barriers: Family	0.82 [0.56, 1.21]	2.39 [0.85, 6.67]	1.60 [0.89, 2.89]	1.39 [0.53, 3.65]	0.50 [0.30, 0.84]**
Job-related stigma, scale of 1–5	1.02 [0.88, 1.19]	1.53 [0.95, 2.45]	0.90 [0.71, 1.13]	0.77 [0.55, 1.10]	0.63 [0.53, 0.76]***
Need for secrecy, scale of 1–5	0.84 [0.70, 1.01]	0.89 [0.58, 1.36]	0.94 [0.72, 1.24]	0.76 [0.49, 1.16]	0.86 [0.67, 1.11]
Anyone close deployed to Iraq War	0.65 [0.44, 0.97]*	0.65 [0.26, 1.62]	0.84 [0.46, 1.52]	0.31 [0.11, 0.87]*	0.51 [0.29, 0.88]*
Received mental health care due to Iraq War[c]	0.79 [0.49, 1.28]	0.82 [0.27, 2.52]	1.12 [0.57, 2.20]	6.89 [1.70, 27.93]**	0.65 [0.35, 1.22]

NOTES: *p<.05, ** p<.01, ***p<.001. Percentages may not add to 100 due to rounding.
[a]Satisfaction is represented by five selected HEDIS indicators from the ECHO survey items.
[b]Among those who received counseling, treatment, or medication in the past six months for a mental health problem.
[c]Among those reporting deployment.

Administrative Claims Data: Supplemental Data Tables

The tables in this appendix provide in-depth data from administrative claims records on TRICARE mental health services users and non-users. These tables serve to supplement the discussion and data presented in Chapter Five.

Table E.1
Data on Eligible Beneficiaries, by Demonstration Area

| | Demonstration Area | | | | | | | |
| | Total Demonstration Areas | | Ft. Carson | | Offutt AFB | | USAF Academy | |
	2002	2003	2002	2003	2002	2003	2002	2003
Total number of eligible beneficiaries (18 years old or older)[a]	134,616	137,187	46,967	48,673	34,653	35,793	52,996	52,721
Total number who meet inclusion criteria	12,462	13,876	4,457	5,178	3,309	3,633	4,696	5,065
Total eligible beneficiaries (adjusted)[b]	149,327	152,179	52,100	53,992	38,440	39,704	58,787	58,482
Percentage of beneficiaries by inclusion criteria (adjusted)								
Saw a mental health provider	2.8%	2.9%	2.6%	2.9%	2.8%	2.8%	2.9%	3.0%
Had a mental health diagnosis	6.8%	7.4%	7.2%	8.1%	6.5%	7.0%	6.5%	6.9%
Received a psychotropic medication	5.7%	6.1%	5.7%	6.2%	6.1%	6.3%	5.5%	5.9%
Received a mental health procedures (CPT codes)	1.0%	1.2%	1.0%	1.1%	0.8%	1.1%	1.1%	1.3%
One or more criterion	8.4%	9.1%	8.6%	9.6%	8.6%	9.2%	8.0%	8.7%
Total mental health users who saw a mental health provider during the year	4,180	4,394	1,378	1,537	1,086	1,108	1,716	1,749
Licensed mental health counselor	14.4%	17.1%	11.5%	15.5%	26.7%	27.5%	9.0%	11.8%
Other mental health providers								
Psychologist	25.9%	24.8%	24.8%	26.5%	23.9%	21.4%	28.1%	25.4%
Clinical social worker	26.2%	23.4%	26.0%	23.2%	28.0%	26.3%	25.3%	21.7%
Marriage and family therapist	11.7%	11.9%	14.4%	14.8%	1.9%	1.4%	15.7%	16.0%
Psychiatric nurse practitioner	5.4%	7.2%	7.5%	10.6%	2.3%	3.2%	5.7%	6.8%
Physician								
Psychiatrist	42.5%	45.2%	39.3%	39.3%	52.8%	56.6%	38.6%	43.3%
Total mental health services users who did not see a mental health provider	8,282	9,482	3,079	3,641	2,223	2,525	2,980	3,316

NOTES: 2002 represents the pre-demonstration year; 2003 represents the post-demonstration year; percentages may not add to 100 due to rounding.
[a] Based on DEERS data as of April 30 of study year.
[b] Adjusted for missing users in April 30 DEERS data.

Table E.2
Data on Eligible Beneficiaries, by Non-Demonstration Area

	Non-Demonstration Area							
	Total Non-Demonstration Areas		Ft. Hood		Luke AFB		Wright-Patterson AFB	
	2002	2003	2002	2003	2002	2003	2002	2003
Total number of eligible beneficiaries (18 years old or older)[a]	208,770	215,794	100,431	101,574	68,702	72,328	39,637	41,892
Total number who meet inclusion criteria	19,965	22,154	7,635	8,525	9,296	10,343	3,034	3,286
Total eligible beneficiaries (adjusted)[b]	231,584	239,376	111,406	112,674	76,210	80,232	43,969	46,470
Percent by inclusion criteria (adjusted)								
Saw a mental health provider	2.3%	2.4%	2.4%	2.6%	2.3%	2.5%	1.9%	1.7%
Had a mental health diagnosis	6.4%	7.0%	5.7%	6.4%	8.0%	8.7%	5.4%	5.8%
Received a psychotropic medication	6.0%	6.3%	4.7%	4.9%	8.8%	9.4%	4.6%	4.7%
Received a mental health procedures (CPT codes)	0.6%	0.7%	0.7%	0.8%	0.5%	0.7%	0.5%	0.5%
One or more criterion	8.7%	9.3%	6.9%	7.6%	12.2%	12.9%	6.9%	7.0%
Total mental health users who saw a mental health provider during the year	5,369	5,778	2,734	2,990	1,815	1,995	820	793
LMHC	11.1%	12.1%	9.7%	12.0%	8.7%	8.9%	21.1%	20.7%
Other Mental Health Providers								
Psychologist	23.4%	24.5%	21.3%	21.6%	23.7%	25.7%	30.2%	32.0%
Clinical social worker	27.4%	25.8%	35.1%	32.1%	21.2%	21.1%	15.5%	13.7%
Marriage and family therapist	6.5%	5.8%	9.6%	8.2%	4.3%	4.1%	0.7%	0.9%
Psychiatric nurse practitioner	2.5%	3.5%	2.8%	4.8%	2.5%	2.7%	1.5%	1.0%
Physician								
Psychiatrist	56.9%	55.5%	54.7%	55.8%	62.5%	57.1%	52.0%	50.1%
Total mental health users who did *not* see a mental health provider during the year	14,596	16,376	4,901	5,535	7,481	8,348	2,214	2,493

NOTES: 2002 represents the pre-demonstration year; 2003 represents the post-demonstration year; percentages may not add to 100 due to rounding.
[a] Based on DEERS data as of April 30 of study year.
[b] Adjusted for missing users in April 30 DEERS data.

Table E.3
Demographic Characteristics of Mental Health Care Users and Non–Mental Health Care Users, by Demonstration and Non-Demonstration Areas and Year

	Demonstration Areas								Non-Demonstration Areas							
	Users				Non-Users				Users				Non Users			
	2002		2003		2002		2003		2002		2003		2002		2003	
	N	%	N	%	N	%	N	%	N	%	N	%	N	%	N	%
Number (N) of beneficiaries	12,462		13,876		135,503		136,786		19,965		22,154		209,438		214,801	
Gender																
Female	8,472	68.0%	9,453	68.1%	56,527	46.3%	57,075	46.3%	13,917	69.7%	15,469	69.8%	87,744	46.5%	89,605	46.3%
Male	3,988	32.0%	4,423	31.9%	65,609	53.7%	66,219	53.7%	6,046	30.3%	6,683	30.2%	101,023	53.5%	103,976	53.7%
Missing/unknown	0	0.0%	0	0.0%	18	0.0%	17	0.0%	2	0.0%	2	0.0%	38	0.0%	59	0.0%
Race																
White	1,956	15.7%	2,316	16.7%	45,935	37.6%	47,095	38.2%	2,263	11.3%	2,815	12.7%	56,656	30.0%	61,188	31.6%
Black	244	2.0%	312	2.2%	7,463	6.1%	7,672	6.2%	438	2.2%	495	2.2%	20,682	11.0%	21,172	10.9%
Other	122	1.0%	141	1.0%	4,513	3.7%	4,535	3.7%	207	1.0%	220	1.0%	8,302	4.4%	8,737	4.5%
Missing/unknown	10,140	81.4%	11,107	80.0%	64,243	52.6%	64,009	51.9%	17,057	85.4%	18,624	84.1%	103,165	54.6%	102,543	53.0%
Marital status																
Married	3,074	24.7%	3,535	25.5%	49,363	40.4%	50,455	40.9%	4,861	24.3%	5,492	24.8%	77,168	40.9%	79,925	41.3%
Divorced	206	1.7%	253	1.8%	3,289	2.7%	3,520	2.9%	305	1.5%	363	1.6%	5,730	3.0%	6,349	3.3%
Separated/annulled	2	0.0%	2	0.0%	38	0.0%	66	0.1%	7	0.0%	4	0.0%	84	0.0%	81	0.0%
Never Married	340	2.7%	349	2.5%	14,836	12.1%	14,899	12.1%	377	1.9%	421	1.9%	21,951	11.6%	22,624	11.7%
Widow/widower	66	0.5%	80	0.6%	379	0.3%	457	0.4%	118	0.6%	109	0.5%	646	0.3%	784	0.4%
Missing/unknown	8,772	70.4%	9,657	69.6%	54,249	44.4%	53,914	43.7%	14,297	71.6%	15,765	71.2%	83,226	44.1%	83,877	43.3%
Member category																
Active duty	594	4.8%	585	4.2%	29,652	24.3%	29,764	24.1%	540	2.7%	573	2.6%	51,949	27.5%	52,499	27.1%
Active duty dependent	2,326	18.7%	2,663	19.2%	18,018	14.8%	18,089	14.7%	3,360	16.8%	3,695	16.7%	29,009	15.4%	29,174	15.1%
Retired	2,897	23.2%	3,274	23.6%	30,388	24.9%	30,510	24.7%	4,786	24.0%	5,387	24.3%	48,835	25.9%	48,913	25.3%
Retiree dependent	5,162	41.4%	5,727	41.3%	35,593	29.1%	35,822	29.1%	8,889	44.5%	9,891	44.6%	53,671	28.4%	54,022	27.9%
Academy student	22	0.2%	22	0.2%	4,495	3.7%	4,280	3.5%	0	0.0%	0	0.0%	44	0.0%	44	0.0%

Table E.3—Continued

| | Demonstration Areas | | | | Non-Demonstration Areas | | | |
| | Users | | Non-Users | | Users | | Non Users | |
	2002	2003	2002	2003	2002	2003	2002	2003
Other	213 / 1.7%	327 / 2.4%	3,993 / 3.3%	4,809 / 3.9%	316 / 1.6%	464 / 2.1%	5,271 / 2.8%	8,927 / 4.6%
Missing	1,248 / 10.0%	1,278 / 9.2%	15 / 0.0%	37 / 0.0%	2,074 / 10.4%	2,144 / 9.7%	26 / 0.0%	61 / 0.0%
Sponsor's branch of service								
Army	4,516 / 36.2%	5,295 / 38.2%	47,179 / 38.6%	49,491 / 40.1%	8,659 / 43.4%	9,757 / 44.0%	107,096 / 56.7%	109,295 / 56.4%
Air Force	5,701 / 45.7%	6,172 / 44.5%	66,916 / 54.8%	65,553 / 53.2%	6,650 / 33.3%	7,357 / 33.2%	64,719 / 34.3%	66,374 / 34.3%
Navy (includes Navy afloat)	755 / 6.1%	865 / 6.2%	6,328 / 5.2%	6,407 / 5.2%	1,886 / 9.4%	2,073 / 9.4%	11,930 / 6.3%	12,214 / 6.3%
Marine Corps	191 / 1.5%	212 / 1.5%	1,365 / 1.1%	1,431 / 1.2%	513 / 2.6%	588 / 2.7%	3667 / 1.9%	4,309 / 2.2%
Other	1,299 / 10.4%	1,332 / 9.6%	366 / 0.3%	429 / 0.3%	2,257 / 11.3%	2,379 / 10.7%	1,393 / 0.7%	1,448 / 0.7%
Age								
18–24	1,598 / 12.8%	1,774 / 12.8%	26,218 / 21.5%	26,799 / 21.7%	2,089 / 10.5%	2,258 / 10.2%	40,597 / 21.5%	41,794 / 21.6%
25–34	1,467 / 11.8%	1,778 / 12.8%	19,540 / 16.0%	19,692 / 16.0%	2,228 / 11.2%	2,469 / 11.1%	32,825 / 17.4%	34,273 / 17.7%
35–44	1,948 / 15.6%	2,064 / 14.9%	22,029 / 18.0%	21,392 / 17.3%	2,508 / 12.6%	2,696 / 12.2%	29,225 / 15.5%	29,546 / 15.3%
45–54	2,108 / 16.9%	2,306 / 16.6%	20,339 / 16.7%	20,421 / 16.6%	2,972 / 14.9%	3,301 / 14.9%	27,817 / 14.7%	27,960 / 14.4%
55–64	1,724 / 13.8%	1,954 / 14.1%	15,783 / 12.9%	16,047 / 13.0%	3,020 / 15.1%	3,433 / 15.5%	24,641 / 13.1%	25,225 / 13.0%
65 and over	3,617 / 29.0%	4,000 / 28.8%	18,245 / 14.9%	18,960 / 15.4%	7,148 / 35.8%	7,997 / 36.1%	33,700 / 17.8%	34,842 / 18.0%

Table E.4
Characteristics of Mental Health Care Users by Demonstration and Non-Demonstration Areas, Provider Group, and Year

	Demonstration Areas								Non-Demonstration Areas							
	LMHCs		OMH Providers		Psychiatrists		Other Physicians		LMHCs		OMH Providers		Psychiatrists		Other Physicians	
	2002	2003	2002	2003	2002	2003	2002	2003	2002	2003	2002	2003	2002	2003	2002	2003
Number of mental health care users	603	750	2,050	1,897	1,527	1,747	8,282	9,482	595	700	1,959	2,160	2,815	2,918	14,596	16,376
Percentage by gender																
Female	83.3%	80.3%	77.3%	76.5%	74.0%	74.1%	63.5%	64.4%	80.8%	82.7%	80.7%	78.5%	79.4%	78.9%	65.9%	66.5%
Male	16.8%	19.7%	22.7%	23.5%	26.0%	25.9%	36.5%	35.6%	19.2%	17.3%	19.3%	21.5%	20.6%	21.1%	34.1%	33.5%
Percentage by race																
White	16.1%	18.4%	18.7%	20.5%	15.8%	16.4%	14.9%	15.9%	17.6%	18.7%	15.6%	17.5%	12.0%	14.1%	10.4%	11.6%
Black	2.2%	2.5%	2.1%	2.6%	1.4%	1.4%	2.0%	2.3%	3.0%	3.7%	4.2%	3.8%	2.0%	2.1%	1.9%	2.0%
Other	1.0%	0.9%	1.5%	1.3%	0.7%	0.9%	0.9%	1.0%	1.5%	2.9%	2.2%	2.1%	1.3%	1.3%	0.8%	0.7%
Missing	80.8%	78.1%	77.6%	75.7%	82.1%	81.3%	82.2%	80.8%	77.8%	74.7%	78.0%	76.7%	84.8%	82.6%	86.9%	85.7%
Percentage by member category																
Active duty	3.2%	3.1%	2.2%	1.7%	3.1%	1.5%	5.8%	5.3%	4.2%	5.0%	3.2%	3.4%	1.9%	1.6%	2.7%	2.5%
Active duty dependent	41.3%	40.5%	41.1%	39.7%	27.0%	28.7%	9.9%	11.7%	45.2%	43.3%	44.3%	42.5%	31.8%	33.5%	9.1%	9.1%
Retired	11.3%	11.3%	15.3%	16.3%	15.9%	16.9%	27.4%	27.3%	10.1%	9.3%	12.8%	12.9%	13.4%	14.7%	28.1%	28.2%
Retiree dependent	30.8%	30.3%	28.6%	29.9%	40.1%	39.8%	45.6%	44.7%	27.4%	25.1%	24.8%	25.0%	38.3%	37.7%	49.1%	49.3%
Academy student	0.0%	0.0%	0.1%	0.2%	0.1%	0.1%	0.2%	0.2%	0.0%	0.0%	0.0%	0.0%	0.0%	0.0%	0.0%	0.0%
Other	1.5%	3.3%	2.0%	2.7%	1.8%	1.7%	1.6%	2.3%	3.0%	3.0%	1.8%	3.1%	1.7%	2.1%	1.5%	1.9%
Missing	11.9%	11.5%	10.6%	9.5%	11.9%	11.3%	9.4%	8.6%	10.1%	14.3%	13.1%	13.2%	13.0%	10.4%	9.5%	8.9%
Percentage by age category																
18–24	22.4%	22.0%	18.1%	18.2%	16.6%	16.5%	10.1%	10.3%	23.2%	21.3%	21.4%	20.6%	16.1%	16.2%	7.4%	7.3%
25–34	24.2%	23.7%	23.0%	22.8%	15.3%	17.2%	7.4%	9.1%	26.1%	31.9%	27.9%	26.6%	20.1%	19.6%	6.6%	6.7%
35–44	26.2%	27.3%	26.1%	24.5%	22.1%	20.1%	11.1%	11.0%	24.7%	24.7%	21.8%	21.9%	18.1%	17.8%	9.8%	9.4%
45–54	19.2%	19.1%	17.8%	18.2%	21.5%	21.0%	15.7%	15.3%	14.8%	15.0%	14.7%	14.2%	20.7%	19.3%	13.8%	14.2%
55–64	7.1%	6.9%	8.0%	9.3%	15.5%	15.7%	15.5%	15.3%	9.2%	6.3%	8.5%	10.5%	13.1%	14.3%	16.6%	16.8%
65 and over	0.8%	0.9%	7.0%	6.9%	9.0%	9.5%	40.2%	39.0%	2.0%	0.9%	5.6%	6.2%	11.9%	12.9%	45.8%	45.7%
Percentage by marital status																
Married	13.4%	14.9%	18.8%	19.2%	17.4%	19.4%	28.3%	28.7%	16.3%	15.3%	16.8%	18.5%	14.8%	16.2%	27.5%	27.5%
Divorced	0.8%	1.9%	1.8%	1.7%	2.0%	1.4%	1.6%	1.9%	1.8%	2.1%	2.1%	1.7%	1.4%	1.3%	1.5%	1.7%

Table E.4—Continued

| | Demonstration Areas | | | | | | | | Non-Demonstration Areas | | | | | | | |
| | LMHCs | | OMH Providers | | Psychiatrists | | Other Physicians | | LMHCs | | OMH Providers | | Psychiatrists | | Other Physicians | |
	2002	2003	2002	2003	2002	2003	2002	2003	2002	2003	2002	2003	2002	2003	2002	2003
Separated/annulled	0.0%	0.0%	0.0%	0.1%	0.0%	0.0%	0.0%	0.0%	0.0%	0.0%	0.0%	0.0%	0.0%	0.0%	0.0%	0.0%
Never married	2.5%	2.0%	1.2%	1.8%	2.0%	1.7%	3.2%	2.8%	2.5%	4.1%	2.5%	2.5%	1.8%	2.2%	1.8%	1.7%
Widow/widower	0.2%	0.1%	0.0%	0.1%	0.3%	0.2%	0.7%	0.8%	0.2%	0.0%	0.2%	0.1%	0.3%	0.2%	0.7%	0.6%
Missing/unknown	83.1%	81.1%	78.1%	77.1%	78.3%	77.2%	66.1%	65.8%	79.2%	78.4%	78.4%	77.1%	81.6%	80.1%	68.5%	68.5%
Percentage by sponsor's branch of service																
Army	27.5%	31.2%	36.7%	42.1%	32.1%	33.5%	37.5%	38.8%	46.4%	47.0%	53.7%	52.0%	47.2%	52.0%	41.1%	41.4%
Air Force	51.1%	50.3%	47.1%	42.3%	49.6%	48.8%	44.3%	43.7%	36.3%	31.9%	26.6%	26.3%	30.6%	27.9%	34.6%	35.1%
Navy (includes Navy Afloat)	7.1%	5.7%	4.2%	4.5%	5.0%	4.9%	6.6%	6.9%	4.5%	4.9%	4.4%	5.0%	6.3%	6.2%	10.9%	10.7%
Marine Corps	2.0%	1.1%	1.1%	1.2%	1.0%	1.0%	1.7%	1.7%	2.4%	1.4%	1.8%	2.2%	1.8%	2.2%	2.8%	2.9%
Other	12.3%	11.7%	10.9%	9.9%	12.3%	11.8%	9.8%	9.0%	10.4%	14.9%	13.6%	14.5%	14.1%	11.7%	10.5%	9.9%
Percentage by study inclusion criteria																
Saw a mental health provider	100.0%	100.0%	100.0%	100.0%	100.0%	100.0%	0.0%	0.0%	100.0%	100.0%	100.0%	100.0%	100.0%	100.0%	0.0%	0.0%
Had a primary mental health diagnosis	99.7%	99.7%	99.5%	99.5%	95.9%	96.2%	71.9%	72.5%	100.0%	100.0%	99.3%	99.0%	93.9%	94.9%	66.4%	68.9%
Received a psychotropic	73.3%	62.5%	54.3%	54.1%	86.0%	83.1%	68.5%	67.2%	70.1%	67.0%	46.4%	46.1%	84.8%	83.0%	70.5%	68.9%
Received a mental health procedure	36.2%	36.3%	29.4%	31.3%	38.4%	46.8%	1.1%	1.0%	23.5%	30.9%	21.3%	21.9%	25.1%	27.9%	1.2%	1.2%

NOTES: 2002 represents the pre-demonstration year; 2003 represents the post-demonstration year; percentages may not add to 100 due to rounding.

Table E.5
Clinical Characteristics of Mental Health Care Users by Demonstration and Non-Demonstration Areas, Provider Group, and Year

	Demonstration Areas								Non-Demonstration Areas							
	LMHCs		OMH Providers		Psychiatrists		Other Physicians		LMHCs		OMH Providers		Psychiatrists		Other Physicians	
	2002	2003	2002	2003	2002	2003	2002	2003	2002	2003	2002	2003	2002	2003	2002	2003
Number of mental health care users	603	750	2,050	1,897	1,527	1,747	8,282	9,482	595	700	1,959	2,160	2,815	2,918	14,596	16,376
Percentage by any mental disorder diagnosis (percentage of mental health care users)																
Mood disorder	64.3%	58.9%	38.4%	42.9%	71.3%	73.6%	24.5%	24.9%	58.7%	61.7%	37.7%	39.9%	74.4%	75.6%	24.6%	25.8%
Anxiety disorder	35.2%	30.7%	30.6%	27.9%	38.4%	35.5%	16.5%	16.9%	44.9%	45.6%	27.4%	27.8%	42.0%	42.8%	18.4%	19.1%
Schizophrenia and other psychotic disorder	3.8%	4.9%	1.5%	2.0%	6.2%	7.0%	3.8%	3.0%	2.5%	3.9%	1.4%	1.7%	6.9%	6.7%	3.5%	3.2%
Adjustment disorder	40.5%	44.0%	48.0%	49.3%	18.0%	16.7%	6.1%	5.7%	44.0%	42.4%	56.7%	55.2%	20.5%	19.8%	5.9%	5.6%
Substance use disorder	12.9%	10.4%	6.1%	6.4%	10.7%	12.9%	26.7%	25.9%	8.9%	8.7%	4.5%	6.8%	8.5%	9.4%	16.8%	18.6%
Conduct/attention disorder	3.2%	3.7%	2.5%	2.4%	4.9%	6.8%	0.8%	0.7%	2.7%	3.7%	1.7%	2.0%	5.9%	6.8%	0.6%	0.7%
Personality disorder	7.0%	4.5%	2.2%	2.2%	4.1%	3.1%	0.6%	0.4%	3.2%	4.1%	2.6%	2.7%	4.8%	4.6%	0.8%	0.6%
Other mental disorder	8.3%	8.1%	9.1%	9.6%	7.0%	8.5%	13.5%	15.1%	6.1%	5.4%	6.4%	5.4%	8.6%	8.5%	15.5%	15.2%
Percentage by primary mental disorder diagnoses (percentage of mental health care users)																
Mood disorder	60.2%	54.4%	33.7%	38.3%	68.2%	69.8%	9.7%	9.5%	53.8%	57.1%	32.2%	34.6%	68.6%	69.9%	9.8%	9.1%
Anxiety disorder	23.7%	22.9%	25.4%	22.4%	26.8%	25.5%	6.3%	6.5%	36.1%	35.6%	21.4%	20.2%	26.9%	27.1%	6.0%	6.3%
Schizophrenia and other psychotic disorder	3.2%	4.3%	1.0%	1.3%	5.4%	5.7%	2.1%	1.7%	2.2%	2.6%	1.1%	1.2%	5.5%	5.1%	1.8%	1.8%
Adjustment disorder	33.5%	38.8%	44.4%	46.0%	13.0%	12.7%	2.7%	2.6%	39.0%	36.9%	52.8%	51.7%	14.0%	13.3%	2.7%	2.5%
Substance abuse disorder	6.0%	5.1%	1.5%	1.7%	4.7%	5.2%	3.8%	3.7%	3.5%	4.3%	1.6%	2.8%	3.2%	3.2%	2.6%	3.5%
Conduct/attention disorder	2.2%	2.8%	1.9%	2.0%	2.9%	4.2%	0.4%	0.3%	1.5%	2.1%	1.1%	1.3%	3.5%	4.0%	0.3%	0.3%
Personality disorder	1.5%	1.5%	0.9%	1.1%	0.9%	1.0%	0.1%	0.1%	0.7%	1.4%	1.1%	1.0%	1.2%	1.2%	0.1%	0.1%
Other mental disorder	4.5%	5.1%	6.2%	6.5%	4.5%	5.6%	6.4%	8.4%	4.4%	3.4%	4.5%	3.4%	5.4%	5.4%	7.1%	7.2%

Table E.5—Continued

	Demonstration Areas								Non-Demonstration Areas							
	LMHCs		OMH Providers		Psychiatrists		Other Physicians		LMHCs		OMH Providers		Psychiatrists		Other Physicians	
	2002	2003	2002	2003	2002	2003	2002	2003	2002	2003	2002	2003	2002	2003	2002	2003
Percentage by presence of DSM-IV comorbidities (percentage of mental health care users)																
Presence of Axis I comorbidity	44.3%	38.7%	24.5%	27.0%	38.7%	40.7%	12.2%	11.9%	43.4%	45.1%	25.6%	28.2%	46.5%	48.8%	11.6%	11.7%
Presence of Axis II comorbidity	6.6%	4.3%	1.7%	1.9%	3.8%	3.1%	0.5%	0.3%	3.2%	3.7%	2.0%	2.3%	4.7%	4.5%	0.6%	0.5%
Presence of Axis III comorbidity	44.9%	39.9%	25.2%	27.9%	39.4%	41.5%	12.4%	12.0%	45.0%	46.0%	26.3%	29.2%	47.4%	49.7%	11.8%	11.8%
Presence of psychosocial problems	1.3%	2.7%	2.3%	2.4%	1.4%	1.7%	0.5%	0.5%	3.0%	4.0%	1.7%	2.7%	1.3%	1.7%	0.9%	0.5%

NOTES: 2002 represents the pre-demonstration year; 2003 represents the post-demonstration year; percentages may not add to 100 due to rounding.

Table E.6
Distribution of Treatment Characteristics Among Mental Health Care Users by Demonstration and Non-Demonstration Areas, Provider Group, and Year

| | Demonstration Areas | | | | | | | | Non-Demonstration Areas | | | | | | | |
| | LMHCs | | OMH Providers | | Psychiatrists | | Other Physicians | | LMHCs | | OMH Providers | | Psychiatrists | | Other Physicians | |
	2002	2003	2002	2003	2002	2003	2002	2003	2002	2003	2002	2003	2002	2003	2002	2003
Number of mental health care users	603	750	2,050	1,897	1,527	1,747	8,282	9,482	595	700	1,959	2,160	2,815	2,918	14,596	16,376
Treatment characteristics (percentage of mental health users)																
Receiving psychotherapy, no medication	9.3%	11.5%	13.9%	13.8%	3.5%	4.9%	0.2%	0.1%	6.6%	9.9%	11.8%	12.3%	2.3%	2.6%	0.2%	0.4%
Receiving therapy and medication	27.6%	25.0%	15.5%	17.7%	35.5%	42.5%	0.8%	0.9%	17.3%	22.2%	9.7%	9.6%	23.5%	25.8%	1.0%	0.8%
Medication only	46.4%	37.9%	38.9%	36.6%	51.3%	41.3%	67.7%	66.4%	53.1%	46.0%	36.9%	36.7%	62.1%	57.7%	69.6%	68.1%
Medication use (percentage of mental health users and mean and median per user)																
Receiving any psychotropic	73.3%	62.5%	54.3%	54.1%	86.0%	83.1%	68.5%	67.2%	70.1%	67.0%	46.4%	46.1%	84.8%	83.0%	70.5%	68.9%
One psychotropic per year	23.1%	23.2%	26.6%	26.3%	24.2%	23.7%	38.2%	38.0%	25.9%	25.4%	24.9%	25.3%	22.0%	24.9%	41.5%	41.8%
Two psychotropics per year	17.6%	16.7%	15.7%	14.2%	22.6%	22.4%	18.1%	17.6%	19.3%	16.9%	12.5%	11.8%	24.9%	22.4%	17.8%	17.7%
Three or more psychotropics per year	32.7%	22.7%	12.1%	13.7%	39.2%	37.0%	12.1%	11.5%	24.9%	24.7%	8.9%	9.1%	37.9%	35.6%	11.2%	9.5%
Mean number of psychotropics per user per year	2.01	1.53	1.05	1.05	2.33	2.20	1.19	1.15	1.69	1.65	0.85	0.84	2.29	2.12	1.17	1.10
Median number of psychotropics per user per year	2.0	1.0	1.0	1.0	2.0	2.0	1.0	1.0	1.0	1.0	0	0	2.0	2.0	1.0	1.0
Type of medication use by drug class (percentage of mental health care users with at least one psychotropic Rx per year)																
Antidepressant	95.0%	87.4%	91.7%	90.1%	86.9%	84.2%	76.4%	75.1%	92.3%	91.7%	87.9%	88.9%	89.2%	87.9%	72.4%	73.4%
Antipsychotic	18.8%	20.7%	6.5%	9.6%	28.3%	30.7%	7.3%	7.9%	15.3%	18.6%	5.9%	6.6%	21.3%	23.3%	5.6%	6.4%
Benzodiazepine	37.1%	35.0%	27.0%	32.1%	36.3%	39.2%	41.7%	43.3%	35.0%	34.1%	29.1%	30.9%	43.9%	41.1%	45.3%	42.9%
Other anxiolytic	6.3%	3.2%	3.4%	0.8%	5.1%	2.0%	2.8%	1.2%	4.1%	2.6%	3.1%	2.7%	4.7%	3.9%	2.7%	1.8%
Mood stabilizer	15.6%	17.9%	7.8%	9.1%	22.3%	22.4%	8.1%	7.9%	11.3%	12.4%	6.2%	6.5%	18.3%	18.5%	7.9%	7.2%
Stimulant	3.6%	4.3%	3.0%	1.9%	4.5%	5.2%	1.3%	1.1%	1.9%	4.3%	0.8%	1.7%	5.2%	5.4%	1.0%	0.8%
Anti-substance use medication	0.7%	0.9%	0.2%	0.8%	0.5%	1.0%	0.4%	0.4%	0.5%	0.9%	0.6%	0.3%	0.5%	0.3%	0.3%	0.2%
Other psychotropic	0.0%	0.0%	0.0%	0.0%	0.0%	0.0%	0.0%	0.0%	0.0%	0.0%	0.0%	0.0%	0.0%	0.0%	0.0%	0.0%

Table E.6—Continued

	Demonstration Areas								Non-Demonstration Areas							
	LMHCs		OMH Providers		Psychiatrists		Other Physicians		LMHCs		OMH Providers		Psychiatrists		Other Physicians	
	2002	2003	2002	2003	2002	2003	2002	2003	2002	2003	2002	2003	2002	2003	2002	2003
Type of Medication Use by Drug Class (percentage of all mental health care users and mean and median per user)																
One psychotropic drug class per year	36.3%	32.8%	36.9%	34.6%	38.8%	35.1%	46.8%	46.2%	40.0%	37.7%	34.2%	32.1%	37.2%	37.0%	49.6%	49.6%
Two psychotropic drug class per year	21.4%	19.3%	14.2%	15.7%	28.0%	29.8%	18.1%	17.7%	20.2%	18.9%	9.4%	11.4%	30.2%	29.5%	17.4%	16.6%
Three or more psychotropic drug class per year	15.6%	10.4%	3.3%	3.8%	19.3%	18.1%	3.7%	3.3%	9.9%	10.4%	2.8%	2.6%	17.5%	16.4%	3.5%	2.8%
Mean number of psychotropic classes per year	1.77	1.69	1.40	1.44	1.84	1.85	1.38	1.37	1.60	1.64	1.34	1.38	1.83	1.80	1.35	1.33
Median number of psychotropic classes per year	1.0	1.0	1.0	1.0	1.0	1.0	1.0	1.0	1.0	1.0	0.0	0.0	1.0	1.0	1.0	1.0

NOTES: 2002 represents the pre-demonstration year; 2003 represents the post-demonstration year; percentages may not add to 100 due to rounding.

Table E.7
Description of Service Utilization Among Mental Health Care Users by Demonstration and Non-Demonstration Areas, Provider Group, and Year

	Demonstration Areas								Non-Demonstration Areas							
	LMHCs		OMH Providers		Psychiatrists		Other Physicians		LMHCs		OMH Providers		Psychiatrists		Other Physicians	
	2002	2003	2002	2003	2002	2003	2002	2003	2002	2003	2002	2003	2002	2003	2002	2003
Number of mental health care users	603	750	2,050	1,897	1,527	1,747	8,282	9,482	595	700	1,959	2,160	2,815	2,918	14,596	16,376
Outpatient visits by mental health care users																
Volume per year (total number of mental health visits by mental health care users)	7,847	9,232	16,601	16,324	13,034	15,298	8,405	9,391	6,505	7,564	13,563	15,292	21,556	22,330	14,480	17,712
Mean number of mental health visits per year by mental health care users	13.01	12.31	8.10	8.61	8.54	8.76	1.01	0.99	10.93	10.81	6.92	7.08	7.66	7.65	0.99	1.08
Mean number of mental health visits per month by mental health care users, for months with any visits	2.44	2.44	2.18	2.21	1.94	1.99	1.10	1.06	2.21	2.26	2.01	2.01	1.77	1.78	1.08	1.11
Inpatient mental health use by mental health care users																
Number of mental health care users who had inpatient service use	57	56	96	123	141	186	1,422	1,684	65	82	108	164	400	401	2,467	2,642
Total number of inpatient episodes among mental health care users	76	79	130	175	194	311	1,765	2128	77	122	189	219	663	675	3,883	3,988
Mean number of inpatient mental health care episodes per inpatient users	1.33	1.41	1.35	1.42	1.38	1.67	1.24	1.26	1.18	1.49	1.75	1.34	1.66	1.68	1.57	1.51
Mean number of inpatient mental health care episodes per all mental health users	0.13	0.11	0.06	0.09	0.13	0.18	0.21	0.22	0.13	0.17	0.10	0.10	0.24	0.23	0.27	0.24
Total number of inpatient days	429	528	1,053	1,501	1,377	2,110	18,065	21,684	397	681	1,169	1,465	4,450	4,481	29,319	38,895
Mean number of inpatient days among inpatient mental health care users	7.53	9.43	10.97	12.20	9.77	11.34	12.70	12.88	6.11	8.30	10.82	8.93	11.13	11.17	11.88	14.72
Mean number of inpatient days among all mental health care users	0.71	0.70	0.51	0.79	0.90	1.21	2.18	2.29	0.67	0.97	0.60	0.68	1.58	1.54	2.01	2.38
Mean length of stay for inpatient episodes (in days)	5.64	6.68	8.10	8.58	7.10	6.78	10.24	10.19	5.16	5.58	6.19	6.69	6.71	6.64	7.55	9.75

Table E.7—Continued

	Demonstration Areas								Non-Demonstration Areas							
	LMHCs		OMH Providers		Psychiatrists		Other Physicians		LMHCs		OMH Providers		Psychiatrists		Other Physicians	
	2002	2003	2002	2003	2002	2003	2002	2003	2002	2003	2002	2003	2002	2003	2002	2003
General health care use—outpatient visits to providers																
Volume of health care visits made by mental health care users	9,719	11,654	23,906	24,231	19,376	23,956	77,261	86,382	8,870	10,035	21,259	23,527	35,454	38,199	145,971	169,985
Mean number of health care visits made by mental health care users	16.12	15.54	11.66	12.77	12.69	13.71	9.33	9.11	14.91	14.34	10.85	10.89	12.59	13.09	10.00	10.38
General health care use—inpatient admissions																
Volume of hospital admissions by mental health care users	117	139	377	461	322	486	4,400	5,017	129	180	413	451	1,117	1,099	9,639	9,959
Mean number of hospital admissions by mental health care users	0.19	0.19	0.18	0.24	0.21	0.28	0.53	0.53	0.22	0.26	0.21	0.21	0.40	0.38	0.66	0.61

NOTES: 2002 represents the pre-demonstration year; 2003 represents the post-demonstration year; percentages may not add to 10 due to rounding.

Table E.8
Additional Utilization Data for Mental Health Care Users by Demonstration and Non-Demonstration Areas, Provider Group, and Year

| | Demonstration Areas | | | | | | | | Non-Demonstration Areas | | | | | | | |
| | Counselor | | OMH Providers | | Psychiatrist | | Other Physicians | | Counselor | | OMH Providers | | Psychiatrist | | Other Physicians | |
	2002	2003	2002	2003	2002	2003	2002	2003	2002	2003	2002	2003	2002	2003	2002	2003
Total number of mental health care users	603	750	2,050	1,897	1,527	1,747	8,282	9,482	595	700	1,959	2,160	2,815	2,918	14,596	16,376
Mental health care utilization rate per eligible beneficiary population	0.4%	0.5%	1.4%	1.3%	1.0%	1.2%	5.6%	6.3%	0.3%	0.3%	0.9%	0.9%	1.2%	1.2%	6.4%	6.9%
Outpatient mental health visits by mental health users																
Volume (total number) of mental health visits by mental health care users for the year	7,847	9,232	16,601	16,324	13,034	15,298	8,405	9,391	6,505	7,564	13,563	15,292	21,556	22,330	14,480	17,712
Volume (total number) of mental health visits per month																
January	663	727	1,574	1,422	1,192	1,348	809	777	576	563	1,248	1,249	1,905	2,026	1,165	1,349
February	623	765	1,399	1,300	1,031	1,302	739	749	556	507	1,138	1,147	1,702	1,766	1,067	1,109
March	638	848	1,356	1,298	1,040	1,294	665	839	563	589	1,101	1,185	1,870	1,940	1,154	1,335
April	736	878	1,480	1,504	1,263	1,443	663	829	622	630	1,201	1,328	1,850	2,153	1,318	1,506
May	664	810	1,373	1,382	1,189	1,280	669	849	609	593	1,160	1,314	1,916	2,049	1,240	1,510
June	631	866	1,270	1,268	916	1,280	629	851	530	623	1,081	1,261	1,637	1,931	1,180	1,574
July	626	799	1,453	1,390	1,108	1,348	732	834	541	682	1,089	1,277	1,702	1,915	1,229	1,571
August	663	706	1,372	1,322	1,042	1,228	730	786	546	632	1,161	1,296	1,837	1,655	1,264	1,528
September	660	809	1,369	1,483	1,033	1,367	693	826	481	706	1,145	1,412	1,838	1,933	1,213	1,672
October	742	768	1,472	1,528	1,197	1,345	712	808	579	752	1,229	1,484	2,022	1,870	1,372	1,713
November	645	645	1,215	1,249	1,050	1,035	695	653	487	635	1,033	1,177	1,647	1,499	1,120	1,408
December	556	611	1,268	1,178	973	1,028	669	590	415	652	977	1,162	1,630	1,593	1,158	1,437
Mean number of mental health visits per year by mental health care users	13.01	12.31	8.10	8.61	8.54	8.76	1.01	0.99	10.93	10.81	6.92	7.08	7.66	7.65	0.99	1.08

Table E.8—Continued

| | Demonstration Areas | | | | | | | | Non-Demonstration Areas | | | | | | | |
| | Counselor | | OMH Providers | | Psychiatrist | | Other Physicians | | Counselor | | OMH Providers | | Psychiatrist | | Other Physicians | |
	2002	2003	2002	2003	2002	2003	2002	2003	2002	2003	2002	2003	2002	2003	2002	2003
Mean number of mental health visits per month (mental health care users; total months)	1.08	1.03	0.67	0.72	0.71	0.73	0.08	0.08	0.91	0.90	0.58	0.59	0.64	0.64	0.08	0.09
Mean number of mental health visits per calendar month with any visits (mental health care users)	2.44	2.44	2.18	2.21	1.94	1.99	1.10	1.06	2.21	2.26	2.01	2.01	1.77	1.78	1.08	1.11

NOTES: 2002 represents the pre-demonstration year; 2003 represents the post-demonstration year. Percentages may not add to 100 due to rounding.

Table E.9
Description of Government Expenditures for Mental Health Care Received by Mental Health Care Users, by Demonstration and Non-Demonstration Areas, Provider Group, and Year

	Demonstration Areas								Non-Demonstration Areas							
	LMHCs		OMH Providers		Psychiatrists		Other Physicians		LMHCs		OMH Providers		Psychiatrists		Other Physicians	
	2002	2003	2002	2003	2002	2003	2002	2003	2002	2003	2002	2003	2002	2003	2002	2003
Number of mental health care users	603	750	2,050	1,897	1,527	1,747	8,282	9,482	595	700	1,959	2,160	2,815	2,918	14,596	1,6376
Expenditures for outpatient mental health visits by mental health care users																
Total expenditures (in $thousands)	$484	$562	$982	$1,020	$923	$1,231	$1,574	$1,995	$409	$501	$770	$980	$1,373	$1,515	$1,309	$1,832
Mean expenditure per mental health care user	$802	$749	$479	$538	$605	$705	$190	$210	$688	$716	$393	$454	$488	$519	$090	$112
Expenditures for inpatient mental health care admissions by mental health care users																
Total expenditures (in $thousands)	$423	$450	$571	$1,238	$685	$1,539	$9,203	$10,645	$258	$533	$568	$996	$2,034	$2,236	$10,577	$14,414
Mean expenditure per mental health care user	$702	$600	$279	$653	$448	$881	$1,111	$1,123	$433	$762	$290	$461	$723	$766	$725	$880
Expenditures for mental health care received by mental health care users																
Total expenditures (in $thousands)	$907	$1,012	$1,553	$2,258	$1,608	$2,770	$10,777	$12,639	$667	$1,034	$1,337	$1,976	$3,407	$3,751	$11,886	$16,245
Mean expenditure per mental health care user	$1,504	$1349	$758	$1,190	$1,053	$1,586	$1,301	$1,333	$1,121	$1,478	$683	$915	$1210	$1285	$814	$992
Expenditures for all outpatient health care received by mental health care users																
Total expenditures (in $thousands)	$900	$1,193	$2,560	$3,409	$2,165	$3,378	$12,232	$16,144	$980	$1,032	$2,290	$2,599	$3,956	$4,378	$19,567	$22,716
Mean expenditure per mental health care user	$149	$1591	$1,240	$1,797	$1,418	$1,934	$1,477	$1,703	$1,648	$1,474	$1,169	$1,203	$1,405	$1,500	$1,341	$1,387
Expenditures for all inpatient admissions by mental health care users																
Total expenditures (in $thousands)	$681	$707	$1,337	$2,068	$1,556	$2,267	$14,720	$17,107	$425	$772	$1,148	$1,534	$3,004	$3,222	$19,289	$24,713
Mean expenditure per mental health user	$1,130	$942	$652	$1,090	$757	$1,298	$1,777	$1,804	$714	$1,102	$586	$710	$1,067	$1,104	$1,322	$1,509

NOTE: 2002 represents the pre-demonstration year; 2003 represents the post-demonstration year.

Table E.10
Visits and Payments to Providers by Demonstration and Non-Demonstration Areas, Provider Type, and Year

| | Demonstration Areas | | | | | | | | Non-Demonstration Areas | | | | | | | |
| | LMHCs | | OMH Providers | | Psychiatrists | | Other Physicians | | LMHCs | | OMH Providers | | Psychiatrists | | Other Physicians | |
	2002	2003	2002	2003	2002	2003	2002	2003	2002	2003	2002	2003	2002	2003	2002	2003
Total number of unique beneficiaries seen	603	750	2,714	2,691	1,778	1,988	5,302	6,013	595	700	3,037	3,260	3,056	3,204	8,648	10,289
Visits to providers																
Total number of visits made by mental health care users	5,569	6,405	22,667	24,242	7,465	7,698	10,865	12,363	4,531	5,302	22,450	25,070	11,112	11,155	19,180	22,679
Mean number of visits made per mental health care users	0.45	0.46	1.82	1.75	0.60	0.57	0.87	0.89	0.23	0.24	1.12	1.13	0.56	0.56	0.96	1.02
Mean number of visits made per mental health care user who saw this type of provider	9.2	8.5	8.4	9.0	4.2	4.0	2.0	2.1	7.6	7.6	7.4	7.7	3.6	6.5	2.2	2.2
Payments made to provider (in $)																
Total payments by government to provider	277,872	309,563	1,166,402	1,292,224	423,694	450,060	2,095,507	2,755,740	238,315	299,216	1,246,387	1,528,186	547,406	597,587	1,828,014	2,403,391
Average payment for provider type per mental health are user	461	413	430	480	238	226	395	458	401	427	410	469	179	187	211	234
Average payment per user in this provider group	22.30	22.31	93.60	93.13	34.00	32.43	168.15	198.60	11.94	13.51	62.43	68.98	27.42	26.97	91.56	108.49

NOTE: 2002 represents the pre-demonstration year; 2003 represents the post-demonstration year.

Bibliography

Berelson, B., *Content Analysis in Communication Research*, Glencoe, Ill.: Free Press, 1952.

Bray, R. M., L. L. Hourani, K. L. Rae, J. L. Dever, J. M. Brown, A. A. Vincus et al., *2002 Department of Defense Survey of Health Related Behaviors Among Military Personnel,* Research Triangle Park, N.C.: Research Triangle Institute, RTI/7841/006-FR, 2003.

Daniels A.S., J. A. Shaul, P. Greenburg, and P. D. Cleary, "The Experience of Care and Health Outcomes Survey (ECHO™): A Consumer Survey to Collect Ratings of Treatment, Outcomes, and Plans," in M. E. Maruish, *The Use of Psychological Testing For Treatment Planning and Outcomes Assessment*: *Adult Assessment Instrumentation*, Vol. III, 3rd ed.). Mahwah, N.J.: Lawrence Erlbaum Associates, forthcoming.

DiMatteo, M. R., R. D. Hays, and C. D. Sherbourne, "Adherence to Cancer Regimens: Implications for Treating the Older Patient," *Oncology*, Vol. 6, 1992, pp. 50–57.

DiMatteo, M. R., C. D. Sherbourne, R. D. Hays, et al., "Physicians' Characteristics Influence Patients' Adherence to Medical Treatment: Results from the Medical Outcomes Study," *Health Psychology*, Vol. 12,1993, pp. 93–102.

Donabedian, A., *The Definition of Quality and Approaches to Its Management,* Volumes I–III, Ann Arbor, Mich.: Health Administration Press, 1980, 1982, and 1984.

Dunnigan, J., "Suicide, Iraq, and Things No One Wants to Talk About," StrategyWorld.com Web site, March 30, 2004.

Eisen, S. V., B. Clarridge, V. Stringfellow, J. A. Shaul, and P. D. Cleary, "Toward a National Report Card: Measuring Consumer Experiences with Behavioral Health Services," in B. Dickey and L. Seder, eds., *Achieving Quality in Psychiatric and Substance Abuse Practice: Concepts and Case Reports*, Washington, D.C.: APA Press, 2000.

Eisen, S. V., J. A. Shaul, B. Clarridge, D. Nelson, J. Spink, and P. D. Cleary, "Development of a Consumer Survey for Behavioral Health Services," *Psychiatric Services,* Vol. 50, No. 6, 1999, pp. 793–798.

Hoge, C. W., C. A. Castro, S. C. Messer, D. McGurk, D. I. Cotting, and R. L. Koffman, "Combat Duty in Iraq and Afghanistan, Mental Health Problems, and Barriers to Care," *New England Journal of Medicine,* Vol. 351, No. 1, 2004, pp. 13–22.

"Information Paper: On Independent Practice of Mental Health Counselors Under TRICARE," provided by TMA (Office of the Chief Medical Officer) to Congress, May 2, 2000.

Institute for Defense Analyses, Altarum Institute, and Mathematica Policy Research Institute, *Evaluation of the TRICARE Program, FY2004 Annual Report to Congress,* March 1, 2004.

Krippendorf, K., *Content Analysis: An Introduction to Its Methodology,* Beverly Hills, Calif.: Sage Publications, 1980.

Kroenke, K., R. L. Spitzer, and J. B. Williams, "The PHQ-9: Validity of a Brief Depression Severity Measure," *Journal of General Internal Medicine*, Vol. 16, 2001, pp. 606–613.

Link, B. G., J. Mirotznik, and F. T. Cullen, "The Effectiveness of Stigma Coping Orientations: Can Negative Consequences of Mental Illness Labeling Be Avoided?" *Journal of Health and Social Behavior*, Vol. 32, 1991, pp. 302–320.

The National Defense Authorization Act for Fiscal Year 2001, PL 106-398, approved October 30, 2000, 114 Stat. 1654.

"Notice of a Demonstration Project for Expanded Access to Mental Health Counselors," 67 Fed. Reg. 57, 581, September 11, 2002.

Orasanu, J. M., and P. Backer, "Stress and Military Performance," in J. E. Driskell and E. Salas, eds., *Stress and Human Performance*, Mawhah, N.J.: Lawrence Erlbaum Associates, 1996.

Schone, E. M., H. A. Huskamp, and T. W. Williams, "Factors Affecting Use of Mental Health Services Among an Enrolled Population in the Military Health System," presentation at the 2003 Annual AcademyHealth meeting, Nashville, Tenn., June 2003.

Spitzer, R. L., K. Kroenke, and J. B. Williams, "Validation and Utility of a Self-Report Version of PRIME-MD: The PHQ Primary Care Study: Primary Care Evaluation of Mental Disorders, Patient Health Questionnaire," *Journal of the American Medical Association*, Vol. 282, No. 18, 1999, pp. 1737–1744.

TRICARE Policy Manual 6010.54-M, "Mental Health Counselor," Chapter 11, Section 3.10, U.S. Department of Defense Military Health System, TRICARE Management Activity, August 1, 2002 (http://www.tricare.osd.mil/TP02/C11S3_10.pdf; last accessed February 2005).

Ware, J. E., Jr., and C. D. Sherbourne, "The MOS [Medical Outcomes Study Medical Outcomes Study] 36-Item Short-Form Health Survey (SF-36), I. Conceptual Framework and Item Selection," *Medical Care*, Vol. 30, 1992, pp. 473–483.

Weber, R. P., *Basic Content Analysis*, Newbury Park, Calif.: Sage Publications, 1990.

Wells, K. B., C. Sherbourne, M. Schoenbaum, N. Duan, L. Meredith, J. Unützer, J. Miranda, M. F. Carney, and L. V. Rubenstein, "Impact of Disseminating Quality Improvement Programs for Depression in Managed Primary Care: A Randomized Controlled Trial," *Journal of the American Medical Association*, Vol. 283, 2000, pp. 212–220.

Williams, T. V., "The Military Health System: Assessment of Substance Abuse and Mental Health," presentation to Substance Abuse and Mental Health Services Administration, Rockville, Md., October 13, 2003.